For my parents,
James and Iva Lou Bullard

An Environmental Journey
Tales of Tragedy and Hope from the Heartland

Published by Backyard Naturalist Press
Springfield, MO 65804

Graphic Design by Hannah Rowlett
Text editing by Rachel Kersey

Publisher's Cataloging-in-Publication
Names: Bullard, Loring, author.
Title: An environmental journey : tales of tragedy and hope from the heartland / Loring Bullard.
Description: Springfield, MO : Backyard Naturalist Press, [2026] | Includes bibliographical references and index.
Identifiers: LCCN: 2026903398 | ISBN: 9798994772119
Subjects: LCSH: Environmental protection--Missouri. | Environmental degradation--Missouri. | Sustainability--Missouri. | Missouri--Environmental conditions. | BISAC: SCIENCE / Natural History. | SCIENCE / Environmental Science. | SCIENCE / Life Sciences / General.
Classification: LCC: TD171.3.M8 B85 2026 | DDC: 363.7009778--dc23 2026

an Environmental Journey

Tales of Tragedy and Hope from the Heartland

LORING BULLARD

FOREWORD

In the heart of America, Missouri stands as a testament to the intricate dance between nature and humanity—a landscape rich in resources and culture, with a complex environmental history. It is a land of rivers and rolling hills, of lush forests and fertile plains, but it is also a land marked by the footprints of industry and the scars of exploitation. As we embark on the enlightening journey through the pages of this book, we are reminded of the stories that shape our relationship with the natural world and how they should inform our perspective of the future.

Loring Bullard is not merely a chronicler of events; he is a passionate conservationist and a steadfast Missouri environmental warrior who has dedicated his life to the cause of natural resource stewardship. His profound understanding of Missouri's ecological past is matched by an undiminished optimism for its future. In this foreword, I invite you to join me in reflecting upon the rich narratives woven into the fabric of our state's environmental history—narratives that encompass both our collective struggles and the triumphs we have achieved together.

Missouri's environmental story is one of contradictions and complexities. It is a tale that runs deep into our state's roots, beginning with the indigenous peoples who cared for the land long before settlers arrived. Their profound respect for nature and sustainable practices serves as a poignant reminder of the harmony that can exist between humanity and the environment. Yet, as waves of settlers arrived, driven by the promise of profit and prosperity, the landscape underwent a dramatic transformation. The allure of lead and coal mining, once seen as a pathway to economic growth, left a legacy of contamination and environmental degradation that still haunts us today.

One cannot reflect on Missouri's environmental history without acknowledging the dark chapters associated with uranium processing and atomic weapons production in the years after World War II. This era marked a pivotal moment in our state's narrative, as the race for nuclear supremacy came at a tremendous cost to both our environment and the health of our communities. Bullard does not shy away from confronting these uncomfortable truths. Instead, he illuminates them, urging us to learn from our past mistakes as we strive to forge a more sustainable future.

However, this book is not merely a chronicle of challenges; it is a testament to the ideals of resilience and hope. Bullard's work serves as a clarion call for the younger generations who will inherit the legacy we leave behind. He reminds us that, despite the daunting issues we face today—water quality degradation, climate change, the alarming loss of biodiversity—there is a glimmer of hope in the actions of private landowners and community members. In

Missouri, we have witnessed ordinary citizens rise to the occasion, transforming their own lands into sanctuaries of biodiversity, promoting sustainable practices, and advocating for clean water initiatives. These grassroots movements are the lifeblood of our environmental future, fueled by the conviction that we can—and must—do better.

The strength of a community lies in its ability to come together in times of adversity. Bullard emphasizes this collaborative spirit throughout his narrative, highlighting the power of collective action. From local conservation groups to statewide initiatives, Missourians have repeatedly demonstrated that when we unite, we can confront even the most significant environmental challenges. Our history is replete with stories of individuals who rolled up their sleeves, rallied their neighbors, and fought for the preservation of our natural resources. It is this spirit of camaraderie and collective responsibility that will guide us as we navigate the uncertain waters of the future.

As we confront the multifaceted threats posed by climate change, it becomes increasingly clear that the path forward will not be easy. Yet, Bullard's narrative instills a sense of agency within us. He reminds us that every action, no matter how small, contributes to a larger mosaic of environmental restoration and resilience. Whether it is a farmer implementing sustainable practices, a family conserving water, or a community advocating for cleaner air, each of these efforts adds to the collective strength of our movement.

In this book, Bullard not only chronicles our past; he invites us to participate in shaping our future. He urges us to engage with our own environmental stories, to reflect on the challenges we have faced, and to recognize the strength that lies within our communities. The lessons learned from our environmental history serve as a guiding light, illuminating the path forward for future generations. It is imperative that we pass on this knowledge—of both our victories and our missteps—to those who will inherit our legacy.

As we turn the pages of this Environmental History, I encourage us to embrace the stories that have shaped our state and nation—to acknowledge the complexities of our environmental past while remaining steadfast in our commitment to a sustainable future. Bullard's heartfelt narrative invites us to remember that we are not merely caretakers of the land; we are its stewards, its advocates, and its defenders.

I encourage you to dive deeply into the insights and reflections that follow. Let Bullard's words resonate within you, inspiring you to become an active participant in the ongoing narrative of Missouri's environmental history. Together, we can honor our past, mobilize our communities, and forge a resilient future for the generations yet to come. The journey is challenging, but as history has shown, we possess the strength and resolve to rise to the occasion—together.

Sara Parker Pauley,

former Director, Missouri Department of Conservation
former Director, Missouri Department of Natural Resources

TABLE OF CONTENTS

INTRODUCTION

I was born and raised in Missouri. In my seventy-three years, I have traveled over much of the U.S. and a considerable portion of the globe, venturing to Scotland, Wales, Iceland, Costa Rica, and Peru. Natural and man-made wonders have awed and inspired me—the amazing engineering of Machu Picchu; the cathedrals, castles, and dungeons of Europe; howler monkeys haunting the rain forests of Costa Rica; the black sand beaches of Iceland, where stranded icebergs glow in the subdued sunlight. In retirement, my wife and I have greatly enjoyed visiting remarkable places, meeting new people, and learning about other cultures. But I'm always happy to get back to my home state, and my beloved Ozarks.

Missouri, after all, has a lot to offer. At first glance, it might seem to lack the charisma of other states, located as it is in the center of the continent, far from the big mountains and coasts. But what it lacks in sandy beaches and soaring mountains it more than makes up for in magnificent forests and beautiful rivers. Intertwined with the dramatic expanse of history, the state lies at the confluence of the nation's two largest rivers—the Missouri and the Mississippi—making it an early center of interstate commerce and a launching point for western exploration. Missouri has long been at the crossroads, the intersection of important trails, roads and waterways—from the far north fur trade to the bustling port of New Orleans; from the embarkation point of early settlers in Pittsburgh to the glacial heights of the Rocky Mountains.

Missouri's mid-continent position also makes it ecologically diverse, the meeting place of several major biomes, or regions with similar climatic and vegetative conditions. The great deciduous forest biome, covering much of the eastern U.S., advanced westward over Missouri, fingering up the valleys of the prairie grassland biome, which had spread from the north and west. The deserts of the southwest pushed toward southwestern Missouri, expressed as dry glades on rocky southern exposures, and the low-lying Mississippi Embayment entered Missouri from the southeast as the once extensive, trackless swamplands of the Bootheel.

This mixing of ecosystem types created great biologic diversity. Huge catfish and sturgeon plied the big rivers, cruising over beds thickly set with mussels. Voluminous springs, abundant in the southern part of the state, cooled and sustained streams supporting uniquely adapted plants and animals. Rolling prairies, once stretching to the horizon, blazed with wildflowers in summer and swarmed with insect pollinators. Buffalo, elk, bear, and deer roamed the prairies and open savannas and woodlands. Extensive marshes of bald cypress and tupelo gum held millions of wading birds, fish, amphibians, and reptiles. Passenger pigeons and Carolina parakeets once streamed in uncountable numbers across clear blue skies.

This diversity of landscape features and biota provided Native Americans with everything they needed—not just for simple subsistence, but for a high quality of life. They hunted, of course, but also used fire in woodlands to create grassy openings where game would browse. They mined clay for pottery and flint for knives and points. They boiled down mineral spring water to make salt. They cut trees for houses and cleared land for crops, especially corn, beans, and squash. Their activities were widespread and in localized places, intensive, but lasting effects were slight compared to those of the settlers who would displace them.

Most of Missouri's early settlers were of European descent. They came looking for productive farmland in what would eventually become the nation's "bread-basket." Prairies were at first shunned because the tenacious root systems of native grasses were resistant to primitive plows, but with better implements, farmers increasingly sought out deep, fertile prairie soils, with centuries of organic humus under deep-rooted grasses. Prairie lands originally covered most of northern and western Missouri, about one-third of the state. By the 1880s, however, almost all this land had been turned by the plow, so that today less than one percent of the original prairies remain.[1]

In southeast Missouri, waterlogged lands were once extensive. The state had almost five million acres of wetlands, mostly in floodplains of the great rivers. But by the second decade of the twentieth century these "swamps" had been mostly dredged and drained, primarily for logging and agriculture. Today, less than 10% of the state's original wetlands remain.[2] The ecological and hydrological effects of these losses were severe. Few people before modern times realized, for example, that the vast expanse of intact wetlands in the Bootheel had acted like a gigantic sponge, storing huge volumes of floodwater, drastically reducing downstream flooding.

Agriculture, logging and later, urban development dramatically altered the face of Missouri. As a child growing up in Missouri, I had very little appreciation of the changes that people had brought to the landscape. It never occurred to me that things were not always the way I saw them. But as with all children, awareness of outside influences grew over time. In grade school, I became interested in science, especially the natural environment. In fact, I may have been pre-disposed to be an environmentalist, because I followed the logic of Rachel Carson in her book *The Sea Around Us*—her premise that all life is connected, and precious. That book, and her later one, *Silent Spring*, instilled in me a lifelong interest in the environment, and the need to protect it.

In 1965, as a 7th grader, I moved with my family to a little town about fifty miles south of Kansas City. Typical of small Missouri towns, mine had about a dozen businesses lining the main drag—a drug store, bank, barber shop, hardware store, grocery store, lumberyard. Surrounding the town were miles of checkerboard farms, sporting fat cattle, fields of tall green corn and waving heads of milo. Just south of town was the silty, channelized South Grand River. North of town was a Missouri Department of Conservation Wildlife Area, where my dad and I sometimes hunted mushrooms, and a little farther north, a Minuteman ICBM rocket installation. By this time, my interest in science had blossomed, and I was fully aware of what those rockets were, and what would happen if they flew.

Like most kids my age, I spent a lot of time outdoors. In the fall and winter, my friends and I took our 22s to hunt squirrels and rabbits in woodlots and forested ravines, or plink cans and bottles at the city dump. In spring and summer, we loaded our fishing gear into my dad's old station wagon and headed to farm ponds for bluegill and bass fishing. Sometimes, we drove to our favorite campsite on the outskirts of town, an old slough, a cutoff meander of the long-ago channelized South Grand River. We pitched tents and drug minnow seines through the fetid, murky water, roasting the bullheads we caught on a fire built with bulldozed stumps.

Those were mostly fun, carefree times, with little thought given to the future. But I couldn't help but notice that the old slough where we camped, a tiny pocket of "nature" in the abandoned bottomlands, had been mistreated. There were piles of trash dumped at the water's edge. Bulldozers had leveled and scraped away most of the surrounding trees, which lay in heaps like bleaching bones. A few herons speared fish in the far reaches of the slough, but for the most part the birds and animals had disappeared. I wondered why people would do this to the last of the "wild" places. The seeds of environmentalism were already being sown into my nerdy science brain.

From that background of interest and awareness, I became an environmentalist, and have been one ever since. But in the American Midwest, as in many parts of the country, environmentalism was not always a popular pursuit. For much of my working career, the label of "environmentalist" was somewhat of an embarrassment for me. I tried to deflect criticism by calling myself a "conservationist," to be distinguished from those wild-eyed "tree hugger" types. "Tree-hugger," it seemed, had a negative connotation in the "conservative" Midwest.

Why this was so, I never fully understood. Conservation, after all, is defined as protection against loss, and I was all for that, especially when it came to the rivers I liked to float and the forests where I hiked. Trees, after all, were critical to human and environmental health.

Later in life, I paused to reflect on my personal journey as an environmentalist. I realized that the environmental "movement"—the nationwide awakening to the threat of environmental degradation—largely paralleled my own trajectory, from youth to maturity to old age. I literally grew up inside the movement—watched its rambunctious birth; watched it mature; and most recently, it seems, observed it declining in power and enthusiasm. With further reflection in my own more advanced age, I can now take a longer view. I can more fully appreciate the great successes of the movement, while at the same time more objectively judge its shortcomings, especially in the face of the "new" and even greater existential threats looming over us today.

In many ways, the rise of the environmental movement can be tied to the state of world affairs in the early 1960s. The war in Vietnam was heating up. It wasn't just about stopping the spread of Communism; an even darker threat was that the world as we knew it could soon cease to exist. Missouri, in rural areas around my home town, bristled with nuclear missiles, although invisible in their underground

Confluence of the Mississippi and Missouri Rivers. Muddy Missouri on the left *(Midland Daily News)*

The massive Ozark Aquifer underlies much of southern Missouri and provides source waters for Missouri's biggest springs. *(Gayle Harper)*

"silos." Fallout from nuclear weapons' testing in the Nevada desert occasionally drifted over the Missouri countryside. For a kid in grade school in Kansas City, a likely Soviet target, those were scary times. Combined with the struggle for civil rights, the war in Asia, and the growing specter of air and water pollution, the threat of nuclear annihilation brought great anxiety about the future. Young people, especially, pushed the government to stop the war, reduce nuclear weapons, and pass laws to fight pollution, all of which led, directly or indirectly, to the first Earth Day in 1970.

I was eighteen at the time of the first Earth Day, and was largely unaware of it. But the next year, in college, a biology professor, who would soon become a mentor, opened my eyes to the wider world. With his encouragement, I took part in the second Earth Day in April, 1971. From there, I watched as the fledgling environmental movement skyrocketed, pushing government to step up—do the right thing. Before long, the movement was no longer the sole domain of scientists and environmentalists, but had become mainstream, even pulling state and federal lawmakers into the fold. Environmental laws written in the early 1970s achieved a hitherto unheard-of level of bipartisan support. Air and water pollution were addressed first, but over the next few decades a slew of new initiatives burst forth in rapid order to address problems associated with solid waste, hazardous waste, toxics, and mining.

In the decades after the first Earth Day, I reached a sort of professional maturity in the public health and watershed management fields. By then, the movement had also matured, from grassroots activism in its formative years to a full-fledged bureaucracy in adulthood, featuring a whole new generation of organizations and agencies pursuing environmental missions. Newly minted environmental attorneys wrote and enforced thousands of pages of environmental regulations. This time was inspiring for me. I watched with pride how aggressive measures to reduce air, water and land pollution vastly improved the environment.

Blue Spring on the Current River is one of the state's deepest and bluest springs. *(Gayle Harper)*

My first real job, my first stop on the road to professional maturity, was in the environmental health field, where I learned about the effects of toxins and disease germs on human health. In the 1980s and 1990s, environmental agencies faced many challenges in Missouri related to contamination with toxic materials. There was the horrible dioxin problem in Times Beach, resulting in the abandonment of the entire town. As is the nature of things, more and deeper connections with people, places and events were revealed to me with the passage of time. At the time of the Times Beach disaster, for example, I had no known exposure to dioxin. But years later, I learned, somewhat disconcertingly, that I might have been exposed to the chemical during my attendance at a technical school.

I also had no known exposure to radioactive wastes, like those arising from nuclear weapons production facilities in St. Louis and Kansas City. But recently, in researching for this book, I discovered that I most likely was exposed to radiological hazards from nuclear weapons testing in the 1950s. Missouri's lengthy mining past would also catch up with it, involving me only at the periphery. At the epicenter of American lead mining, the state would suffer many ill effects from mining and processing this toxic metal, including high blood lead levels in children living near smelters. I had no personal connection with lead contamination, but while working for the Health Department in Springfield, I investigated potential groundwater contamination from lead mining within five miles of my home.

Coal mining, which was prominent where I grew up in Henry County, left toxic residues in the excavated overburden, the "spoil piles." In high school, I swam in the "strip pits" left over from coal mining, but had no inkling of the environmental impacts represented by those ragged, gaping holes. And today, the burning of coal in Missouri creates a disproportionate share of the nation's air pollution with sulfur dioxide and carbon dioxide gases, prevalent factors in climate change, which affects us all.

In all these cases, rigorous environmental laws passed in the 1970s and 1980s mandated cleanups and regulatory controls that effectively dealt with most of the contamination. The environmental movement had directly led to some of the most significant achievements in pollution reduction ever seen, anywhere on the planet. But the regulatory "environment" had also changed, becoming a battleground for technical experts on both sides—the regulators and the regulated—who argued incessantly over the nuances of standards and regulations. What should be the exact limit of this pollutant in water? What are the exact costs and benefits of this regulation? The world of environmental regulation had become exceedingly complex.

In middle age, I worked in the watershed protection business, navigating this more complicated regulatory world. I could see that water pollution laws from the 1970s had done a remarkable job of cleaning up many of the state's rivers and lakes, but could also see that so-called "non-point source" water pollution, coming from myriad smaller, unregulated sources like septic tanks or runoff from fields and cities, could negatively affect the quality and health of waterways. I worked with farmers on grazing lands to reduce the flow of soil and animal wastes into streams and lakes, but saw how hundreds of animals in confinement could easily overwhelm attempts at controlling water pollution. And living in southern Missouri, I quickly learned about the complexities of karst topography, where pollution at the surface can quickly sink into the subterranean world, contaminating springs or wells many miles away. There was the shocking example from West Plains, where a huge sewage lagoon suddenly funneled into the ground, almost overnight.

Prairie remnant ablaze with summer wildflowers. *(Missouri Department of Conservation)*

While laws generated by the maturing environmental bureaucracy cleaned up pollution, opposition to further laws and regulations also matured, becoming highly organized and politically connected. Business and industry groups, with the backing of many politicians, complained that any newly proposed regulation constituted government over-reach—taking away jobs, stifling the economy. There were even efforts to "roll back" existing regulations, no matter what they were, or how important they were. Increasingly, environmentalism was seen as being too "liberal." As a result, new proposals faced unprecedented levels of political pushback and gridlock.

Most people alive today have no recollections of the 1960s—no firsthand knowledge of how many lakes and rivers were grossly polluted; how much of the air was hazy and dangerous to breathe; how many toxic chemicals were being carelessly "disposed of" in the environment. For the present generation, it has become easy to develop a sense that the worst is far behind us. Our dwindling collective memory of how bad things were has lost its ability to inform and motivate. As a result, many Americans now take their "right" to a clean environment for granted. They don't worry about it, so have little inclination to want more rules to protect it.

Missouri's river bottom and slough wetlands were once extensive.
(Missouri Department of Conservation)

Despite the reach and severity of today's global environmental problems, there are no throngs out there protesting these looming threats; no vocal masses marching in the streets, demanding change. Armies for the environment, so conspicuous in 1970, have largely dissipated. This is unfortunate, because the environmental perils we face today are bigger, and potentially more severe, than the ones we addressed in the 1960s. It could be argued that in its old age, sixty-years plus, the environmental movement has lost much of its strength, and much of its following.

With environmentalists playing only a minor role, there is little public discussion today about the potential effects of environmental degradation on future generations. This is shameful. Providing our progeny with a clean and healthy environment should be among our highest priorities. If we so desire, we have the tools at our disposal to secure a high quality of life for future generations. It is patently unfair to leave for them the messes we now know how to prevent. After all, making life better for our progeny is the one thing that should unite us all as human beings on this remote outpost we call planet Earth.

In the early 20th century, Missouri's wetlands were extensively ditched and drained. *(Glaub Farm Management)*

Now that I'm retired, I think more about the future. I had imagined myself in retirement dwelling in a fairytale land of comfort and ease. That was my plan—to sit back and enjoy life; maybe do some traveling, see the world, but spend most of my time at home, relaxing, puttering in the yard and garden. But when I see or read about today's complex environmental problems—the climate crisis, the loss of biological diversity, the depletion of healthy soil, the pollution of rivers and streams by industrial agriculture—and see the lack of action, and the large measure of indifference, I feel I can't sit by, doing nothing. With daughters and now grandkids, I have a bigger stake in the future. So, I have decided to get back into the fray—do some work on behalf of environmental organizations, and to write.

The huge lumber mill at Grandin in the eastern Ozarks operated from the late 1880s to about 1910. *(State Hist. Society)*

As to the anticipated impacts of my writing, I realize that I can't set my expectations too high—especially now. We have just gone through a tumultuous election year. With social issues so prominently on the minds of American voters, providing a clean and livable environment seemingly slipped to a lower priority. It pains me that the "environment" has been pushed into the "liberal" basket of issues, when it is so clearly about conserving—protecting—our homes, our planet, and our future. But our focus must be on the future, not on the past. Our environmental problems will have to be addressed, somehow, for the sake of our children and grandchildren.

Early lead mining in Washington County. Portion of the Oscar Berninghaus mural in the capitol building in Jefferson City. *(State Historical Society)*

Webb City Mining District, Jasper County, ca: 1902 *(State Historical Society)*

It's easy to become worried about the future, but at the same time, there are reasons for optimism. During my career, I have had the pleasure of interacting with hundreds of people who care deeply about the environment, and who are working tirelessly to make a positive difference. Resource professionals today typically have a deeper ecological understanding than their predecessors. They are cross-pollinating more with others in allied fields, coming out of their silos, approaching problems in an inter-disciplinary way, mirroring the connectedness and synergy of nature. New ideas in long-term land protection are being tested, leading to a "landscape of hope," as explained in my last chapter.

There are yet more reasons for hope. Serving on the boards of environmental organizations, I'm continually exposed to people with positive outlooks—people who know how to work across cultural and political lines. I have seen firsthand how selfless collaboration can result in forging effective and lasting solutions to our environmental problems, even as the complexity and severity of the threats continue to grow. I have seen how more and more people embrace the idea that we all have a responsibility to protect nature on its own terms, on our own land, and for our own good—to turn a land ethic into practical action. These are good trends, creating light at the edge of darkness.

I don't expect to change the world with this book. But I do hope this telling of our state's environmental history, examined through the lens of my firsthand, if minor, experiences, will help keep us from repeating the mistakes of the past. It is helpful, I think, to be reminded of what can happen, but also to be reassured by the fact that over fifty years ago, when environmental threats seemed overwhelming, we somehow were able to come together—debate, compromise, and reach across the political aisles to resolve our environmental problems. The story of our sometimes faltering, sometimes enlightened past should help to remind politicians and others that in the final analysis, protecting the environment is really a human issue, not a political one.

SOURCES

1 Missouri Prairie Foundation website: moprairie.org/mission/outreach-and-education/what-is-prairie/

2 Jane E. Epperson, *Missouri Wetlands: A Vanishing Resource*, Missouri Department of Natural Resources, Water Resources Report No. 39, Rolla, 1992, p. 1

CHAPTER 1: THE HUMAN YARD

"The only possible guarantee of the future is responsible behavior in the present."

WENDELL BERRY

Like many people, I had only vague notions of what I would be doing in retirement. It seemed like rest was called for, and relaxation, and I imagined myself sitting on my back deck, cold beer in hand, watching the birds, bees and butterflies go by. In fact, I looked forward to that aspect of retirement so much that I planted a variety of native plants in hopes of attracting those birds, bees and butterflies. It seemed to me that a yard filled with native plants and buzzing bees would be interesting as well as relaxing. I could get my healthy dose of nature, my biotic fix, right at home.

I also wanted to grow a bit of our own food, so I planted a garden; more as a place to piddle than a significant source of nutrition. I built raised beds for our summer bounty of beets, peppers, strawberries and, most enticingly, those delicious, vine-ripened tomatoes. In one corner of the yard, I installed a little pond with a "creek" cascading into it over a miniature waterfall. I put some aquatic plants in the pond, providing shade and cover for a handful of goldfish. Now it was time to sit back and relax and enjoy my "golden years."

But it was not to be. Early mornings and late evenings were nice, but during the day, in warmer weather, I was frequently disturbed from my afternoon dreamy state by the incessant droning of lawnmowers, the buzzing of weed eaters, and worse yet, the high-pitched whining of leaf blowers. Almost every day, from spring through late fall, a leaf blower started up, somewhere. They were like giant mosquitoes humming right next to my eardrum. I tried to ignore them, but it was impossible. Despite my careful preparation of a comfortable space, urban irritants were undermining my plan for a peaceful retirement. I didn't want to be that grumpy old curmudgeon neighbor, but the auditory insults were getting under my skin.

"Pond" in Author's Yard

For me, it was mainly the noise, but as I investigated further, it became apparent that leaf blowers, especially, had other faults. For one thing, they kick up billowing clouds of dust. Neighbors with asthma or allergy problems would be well advised to go inside or don dust masks. Along with swirling leaf fragments and dust there are pollen grains, mold spores, microbes, dried feces, and lawn chemicals. Many communities began asking if all this blowing is necessary, or even harmful. As a result of rising concerns about health impacts and noise pollution, over 200 cities in the U.S. now have restrictions on gas-powered leaf blowers, often limiting them to certain times of the year.[1]

Whenever a leaf-blower started up nearby, I retreated into the house. I sometimes watched TV, but soon became disgusted with the anti-environmental diatribes of many politicians and their promises to "roll back" regulations. I had spent my entire career working on local solutions to local environmental problems and was incensed to hear state politicians arguing that counties or cities should have no right to pass rules more stringent than the state's. I found myself being drawn out of retirement and back into the fray.

Getting back involved wasn't a bad thing. I felt that I still had something to give to the community, based on my years of service in the public health and watershed protection fields. I found myself allied with people I enjoyed being around, because they were vigorously pursuing what I considered an attribute in short supply—thoughtful stewardship. At the same time, however, I found that my relaxing times at home, in my backyard haven, were slipping away.

That was also okay. My yard, after all, was secondary to my personal fulfillment—and, of course, to my relationships with my wife, my family, and my friends. After all, the yard was still out there, doing its thing, alive with bees, birds and butterflies, who could take care of themselves. The wildflowers had crowded and jostled together, becoming a thick tangle defying interference from "weeds." My garden still produced, although a little haggard looking from lessened attention.

About ten years after I had "officially" retired, I read a book by Doug Tallamy called *Nature's Best Hope*. I later told people that when I read this book, the "scales fell from my eyes," to borrow a Biblical expression. The book brought home to me a critical point—what I did in my yard was every bit as important, maybe even more important, than my part-time involvement in environmental organizations and causes. One might not expect to be inspired by an entomologist, but Tallamy's clever explanations of intricate and surprising ecological relationships blew me away. More importantly, it was a call to arms for urban or suburban dwellers. My wife, Beth, and I began to think in terms of how we could, and should, join a new movement to save the planet one yard at a time.

Tallamy's main point is that we'll never save the planet, or, more accurately, save the planet's ecosystems, unless we take a different view of nature and our place in it. To an alarming extent, we don't teach our children that plants and animals create the life support systems that keep us all alive, or that plants generate the oxygen we breathe, or that animals provide us with pest control services and pollinate almost all plants. "Our ignorance of nature has led us to a dangerous indifference about its fate," Tallamy laments, and ecological enlightenment is the only option left if we want to "remain viable in the future."[2]

Tallamy's books are not about how bad things are, but about hope. He doesn't dwell on the "pox" that has arisen because of our ecological ignorance, but instead talks of a "cure" for it—one that requires only small efforts by many people, but one that will result in enormous physical, psychological and environmental benefits for us all. We can and should begin this long journey right on our own land, where we have the most interest and ultimate control, because "we no longer have the right to ignore the stewardship responsibilities attached to land ownership." We should start small and think ecologically, but expansively, because our land and the ecosystems upon it are essential to everyone's well-being, not just our own.

Best of all is Tallamy's message of inspiration. Homeowners, he says, can derive joy and satisfaction from their individual efforts by "indulging rather than sacrificing." Instead of being asked to give up something (the usual haranguing of environmentalists), we can, by cultivating our own little biotic preserves, add something to the world's beauty and ecological stability—take a proactive approach that Tallamy hopes will eventually become a "culturally embraced imperative."[3]

In the past, we protected land for many reasons, including ecological ones. The national parks, for example, were set aside mainly for their rugged beauty or historical significance, but were also intended to protect threatened or unique plants and animals. Franklin D. Roosevelt added millions of acres of wildlife refuges, largely to provide feeding and resting stops for diminished flocks of waterfowl. Even the United Nations designated certain lands as "places of ecological significance." Tallamy, however, challenges this notion, saying "all places have ecological significance, not just the few places we have not yet destroyed."[4]

He then patiently explains the simple things you can do on your property to start the ecological healing process, of which we can all play a part, because effective conservation isn't beyond the reach of the individual. In fact, it is our efforts as individuals that will determine whether, as a society, we succeed or fail. Rather than waiting for more lands to be set aside for nature preserves, or for conservationists to carry out restoration projects, we can start solving the world's ecological problems right now, right in our own back yards. In fact, we can't afford to leave all these chores to professional conservationists, because there are just too few of them and too little time.

Urban and suburban people will have a big part to play. In the past, it was the rural areas, the farms, which provided refuge for much of the life that shunned the cities. Almost every farm had significant unfarmed sections, like brushy fencerows, patches of timber or pockets of wetlands, which were left to their own, becoming havens for a diversity of plant and animal species. These neglected areas offered, in total, an impressive number of ecosystem services like pollination, oxygen production, flood protection and runoff control.

But after the 1960s, when the U.S. Department of Agriculture called on American farmers to maximize production and plant fencerow to fencerow, farmers obediently removed trees, drained wet areas, and took out ragged, weedy fencerows for more space to plant crops. Today, "clean" fields and fencerows are the norm. As a result, way too many farms have become impoverished landscapes dedicated solely to monocultures of crops or grass. Because they had to understand intricate relationships between soil, water, weather, plants, and animals, farmers have been called the original environmentalists. But these days, increasingly squeezed by ever thinner profit margins, they must think mainly of their economic survival, not ecological restoration.

Urban and suburban spaces haven't fared much better. Historically, they have become ecological dead zones. They have spread rapidly across the landscape, gobbling up rural and farming land, until they now cover over 100,000 square miles, an area 1 ½ times the size of Missouri.[5] When a city's nearby and easy areas were built out, developers began encroaching into ecologically sensitive areas such as wetlands, floodplains, and wooded hillsides. Further, urban dwellers largely saw their living spaces as buffers against nature. They used powerful poisons to fight against invading cockroaches, mice, ants, snakes and weeds. They kept the grass short and the bushes closely trimmed because the Health Department warned them against disease-carrying "vectors" like rats and mosquitoes. As a result, many of them were led to believe that a yard with no animal life was desirable and entirely normal. After all, animals belonged in nature, which was out there, somewhere else.

Wildflower plantings attract hummingbirds and pollinators. *(Loring Bullard)*

Typical suburban residential property near Springfield. *(Loring Bullard)*

But with rural areas losing ground, both geographically and biologically, the urban zones, which are still expanding, have a newfound importance. Urban and suburban people have the power to change the ecological picture, if they so desire. More and more of us are living in cities, about 83% of the U.S. population, up from 64% in 1950, so there is potentially a vast army of small-scale ecological warriors.[6] Unlike farmers, who must think about economic "yields" from their properties, urban and suburban residents are mainly interested in quality of life, which often means adding attractive features to their yards or enticing wildlife. They have the luxury of thinking about their yards as biotic preserves and ecological sanctuaries.

There are some big hills to climb in getting people to think ecologically about their yards, however. The iconic image of the perfect yard has been carried over from Europe, dating from a time when rich landowners, eager to show off their nobility, surrounded themselves with frivolous turf rather than edible crops or pastures. Americans still see the immaculate, sterile lawn as the gold standard in suburban landscaping. The move to more biologically enriched landscapes must be as much culturally as environmentally inspired.

In the Springfield area, we have taken the idea of expansive lawns to new heights. Surrounding the city are hundreds of suburban properties, three to five acres in size, almost totally covered with turf, mostly fescue grass. Some social scientists hypothesize that the desire for short grass and open vistas is programmed into our DNA, from times on the African savannas when we needed to see the lions coming. Regardless, these expanses of short-cropped grasses offer precious little biologic diversity or ecosystem benefits or for that matter, not much, these days, in the way of protection from ravenous beasts.

The 40 million acres of turf in the U.S., if gathered into one vast patch, would almost cover the state of Missouri. According to the EPA, Americans collectively spend more than 3 billion hours on lawn care. But it's not just that. There are about 140 million pieces of gas-powered lawn equipment in the country, annually spewing out 30 million tons of carbon dioxide, the primary greenhouse gas. They also produce a disproportionate share of the volatile organic chemical (VOC) emissions (17%), as well as 21,800 tons of fine particulates, the ones that lodge deep in the lungs.[7] Two-stroke gasoline engines are particularly polluting, releasing as much as one-third of the fuel mix as unburned aerosols. All these emissions contribute to asthma and other respiratory illnesses, and increase risks of stroke, heart failure and arrhythmia.[8]

Then there is all the fertilizer we pour on the grass to make it grow. Homeowners use roughly the same amount of fertilizer on their lawns, in total, as farmers use on their fields. Much of it doesn't stay put. About 40 to 60% of it runs off with rainstorms, polluting streams, or else leaches into groundwater. American homeowners also pour, powder, or spray on an estimated 70 million pounds of pesticides. 40% of these chemicals have been banned in other countries because they are identified as carcinogens. Scores of studies have linked lawn chemicals to cancer risks, especially lymphomas.[9] Children and pets, who spend more time in the grass than adults, are particularly at risk.

And then there are the commonly used landscaping plants, the plug-in "lollipop trees and poodle bushes" in the parlance of one of my eco-evangelist friends. Even Thomas Jefferson, who had a strong hand in our democracy, didn't want common, ordinary American plants on his grounds, but instead opted for European imports as befitting his station in life. It's a tradition that carries on to this day. Some 80% of the plants in American yards evolved in Europe, Asia, or South America.[10] And therein lies a big problem. Because these species are not natives, they are not contributing members of the local ecosystem. They are often selected precisely for their resistance to "pests," meaning that local insects—critical bird food—cannot eat them. They have been chosen for their decorative value, not for any ecosystem values they might possess.

Some of these non-native plants, in fact, have caused serious ecosystem disruptions. The ones that have escaped cultivation and then spread like wildfires throughout urban, suburban, and rural spaces around the country are called invasives. About 85% of the invasive woody plants in the U.S. have escaped from cultivation.[11] These nuisance species crowd out the natives, replacing ecologically rich, stable landscapes with large expanses of impoverished monocultures. Just think kudzu in the south, and bush honeysuckle and wintercreeper in the Midwest.

Instead of exotic ornamentals, and especially obnoxious invasives, we need to use more native plants in landscaping. Clusters of native wildflowers, with their contrasting colors, delicate fragrances and intricate textures, will produce a rich tapestry, an optical and olfactory smorgasbord. But they will also provide a biologic sanctuary for a host of native animals like birds, bees and butterflies. Tallamy suggests we start small. Instead of a complete overhaul of the landscape, start by planting just a few natives in one area, replacing a small patch of turf, for example. In fact, it is helpful to think of turf as the "accent" feature, not the main event, using smaller spaces of mowed lawn to accentuate the beauty of native plantings.

There are many native plants to choose from, and local growers are responding to the rising desire of homeowners to use them. To some extent, the choices are up to individual tastes and preferred size and color palates. But it is a good idea to cater to the animal species that are most imperiled. Planting milkweed to help monarchs is well known, but bees are also in trouble. Half of the native species of Midwestern bees have disappeared over the last century, and a quarter of bumblebee species are at risk of extinction. We also can't forget about bats, hummingbirds and moths, which pollinate widely by day and night. Bats and birds, along with insects, are responsible for pollinating 87% of our plants, including a third of our agricultural crops.[12]

Millions of insects and other invertebrates are found in leaf litter, making it an important food source for birds, mammals, amphibians, and reptiles. *(Xerces Society)*

It's also important to think about trees, and even their leaves. Most urban and suburban lots have just a few trees; so, ecologically speaking, it's important for them to be the right kind of trees. Oak trees, for example, host over 500 species of caterpillars—high-protein bird food—more than any other kind of tree. Native cherry and willow trees are also good. Many of the caterpillars that birds consume develop in leaves on the ground. And what do we typically do with these leaves? We gather them up, bag them, and take them away.

It hit me like a thunderbolt when I read Tallamy's books, because I had never made the connection. We want to attract birds to our yard, but we plant non-native species that host few of the insects that birds need. Even if we have desirable, native trees, we remove the fallen leaves and the insects living in them, one of the most important sources of foods for birds. We try to entice birds to our yards by putting out feeders with seeds, but most baby birds need insect food, not seeds. They need the caterpillars and other insect foods that are mostly found in the leaves and on native plants. By growing non-native trees and other plants, and by removing all the leaves, we cut our yard's ecological potential off at the knees.

In response to that revelation, I have changed how I manage our vast, one-quarter acre estate. The back of our long, narrow lot is shady, almost completely under a canopy of walnut, ash and maple trees. Under the trees, we have created a "no mow" zone where we let the fallen leaves lie. A considerable layer of leaves builds up in the late fall and winter, but amazingly, six or seven months later, they are almost all gone, having decomposed, their sequestered caterpillars eaten or grown up, their organic residues filtered into the soil to provide sustenance for the myriad life forms found there.

To some extent, the clarion call for homeowners to step up has been heard. More and more urban and suburban folks are testing the waters. At least in Springfield, regular outdoor sales of native plants have been heavily patronized. More people than ever are interested in growing and showing their native plants and attracting wildlife to their yards. It may not be a Renaissance moment, but it is a hopeful sign. Maybe we are starting to act a little more "reverently and respectfully" toward nature as Doug Tallamy advises. Maybe we will eventually "stop segregating ourselves from nature and learn to live as part of it."[13]

Environmental Services, a department of the city of Springfield, has launched a "Yard Ethic" program promoting and recognizing environmentally sound yard practices like mulching, composting, soil testing, rainwater harvesting, maintaining a healthy tree canopy, and using native plants in landscaping. The successful petitioner is given a yard sign, which says he or she has achieved a "Yard Ethic" yard. But just as importantly, it announces to passersby, "I care about the Earth." It says to neighbors and strangers alike, "I care about you, because you live here, too." Yard Ethic is in tune with the land ethic espoused by Aldo Leopold, whose careful thinking over many years finally crystallized into his famous observation: "A thing is right when it tends to preserve the integrity, stability, and beauty of the biotic community. It is wrong when it tends otherwise."[14] I especially like that part about beauty, one of the purest pleasures of growing natives.

Yard sign provided by the Missouri Prairie Foundation. *(Loring Bullard)*

Ethics, of course, shouldn't end at the fence. There is the matter of neighborhood ethics, too. Blowing dust and pollen into the air with a leaf blower could affect his or her asthma. Letting your cats run loose in the neighborhood could endanger the birds she is trying to attract to her feeders. The light on your porch shining into his eyes is a constant source of irritation to him, as well as ecologically damaging to nighttime animals. In urban areas, ethics governs close encounters between closely packed neighbors, just as it informs our close-up relationships with the natural world. Positive effects from an ethical person, and his or her yard, will radiate outward, a step toward an ethical neighborhood which, in turn, is a small step toward a more ethical world.

There are other ethical considerations related to yard care and its effects on quality of life. My gas-powered mower or leaf blower adds to the load of greenhouse gases that are causing worsening floods and crop failures for a poor farmer in faraway Bangladesh. What are my ethical responsibilities to him? In the U.S., we spend about $30 billion per year on yard care. According to the World Health Organization, that is about the same amount that

Springfield's Yard Ethic Program recognizes landowners using environmentally beneficial practices. *(Loring Bullard)*

it would take to provide every person on Earth with clean drinking water and basic sanitation.[15] As highly consumptive, image-conscious Americans, what are our ethical responsibilities to the rest of the billions of people who share this planet?

Of course, we can't go around feeling guilty all the time. But we can at least try to take a few positive steps. Maybe we can get our neighbors and yard care companies to switch to battery-powered leaf blowers. Maybe we can look at our own properties and decide to plant a few native wildflowers or quit removing at least some of the leaves. Those are the kinds of things we can feel good about. At the same time, we would be helping our fellow man by not producing as much pollution, or using as many yard chemicals—all while protecting and strengthening our thin and fragile biosphere.

There is ample reason for hope that yard ethics will become more widespread. More and more people are paying attention to ecological health in their yards and gardens. A survey sponsored by the National Wildlife Federation in 2022 indicated that one in four homeowners specifically buys native plants to attract wildlife to their yards.[16] And that number appears to be growing. In the meantime, I can feel better about the time spent puttering in my yard, looking for ways to improve its attractiveness to the birds, bees and butterflies that will comfort me in my golden years.

SOURCES

1 Route-fifty.com/management/2023/10/hear-its-sound-leaf-flower-bans/391543, Oct. 2023

2 Douglas W. Tallamy, *Nature's Best Hope*, Portland, OR, 2022, p. 9

3 *Nature's Best Hope*, p. 12.

4 Ibid., p. 11

5 University of Michigan, Center for Sustainable Systems, css.umich.edu/publications/factsheets/built-environment/us-cities-factsheet

6 Ibid.

7 EPA.gov/banks, "National Emissions from Lawn and Garden Equipment"

8 Barrington.ri.gov, "The Unintended Consequences of Leaf Blowers"

9 *Nature's Best Hope*, p. 48.

10 Ibid., p. 9

11 Ibid., p. 51

12 Ibid., p. 154

13 Ibid., p. 11

14 Soil and Water Conservation Society, *Aldo Leopold: The Man and His Legacy*, 1995, p. 21.

15 World Health Organization, "Global Costs and Benefits of Drinking Water Supply and Interventions to Reach the MDG Target and Universal Coverage," 2015.

16 Nwf.org/Latest-News/Press-Releases/2022/5-02-22-Consumer-Gardening-Report

CHAPTER 2:
A RAIN OF RUIN

"If the radiance of a thousand suns were to burst at once into the sky, that would be like the splendor of the Mighty One; I am become Death, the shatterer of worlds."

BHAGAVAD GITA,

QUOTED BY J. ROBERT OPPENHEIMER UPON THE BOMBING OF HIROSHIMA

When I was a kid, it was the worst thing I could imagine—all out nuclear war—World War Three. Mine was the first generation to live its entire history under the fear of Armageddon; the hellacious fireball, the burning and evaporation of flesh in a monstrous, billowing mushroom cloud. That scenario seemed all too real in late 1962 as I started 5th grade in Kansas City. That October, American air photos revealed that the Soviets were installing offensive missiles in Cuba. A high stakes game of chicken ensued, with President Kennedy ordering a "quarantine" of Cuba, since a blockade was generally recognized as an act of war. Khruschev's missile-laden ships kept steaming toward the Caribbean and for eleven days, the world held its collective breath as neither side would back down.

I was only eleven years old, but I knew full well what a nuclear war would mean. We had duck and cover drills at school nearly every week—filing into hallways, kneeling nose to tail, fingers interlaced behind the head. We watched movies about the horrible destruction of a nuclear blast. Every time I heard a siren, I looked up, half expecting to see a missile streaking across the sky. My heart would pound, my stomach queasy, until I realized the wailing was an ambulance, or a fire truck. We didn't have a fallout shelter at our house, but my dad showed us a tiny room off our concrete basement. It had a musty, sewer smell, but it would do in an emergency, he said, although I couldn't imagine spending two weeks in that dank, dark hole.

I had developed a heightened sense of anxiety about nuclear war even before the Cuban missile crisis. At the Missouri State Fair in Sedalia, my parents marched me past a Civil Defense exhibit—a trailer-mounted diorama depicting the horrifying aftermath of a thermonuclear exchange. On one side was a family, snug and secure and seemingly at ease in their fallout shelter. On the other side were the unfortunates who had not found shelter—bodies sprawled on the ground, burned and blackened. One man had intestines spilling out. If it was meant to install a primal fear, it certainly did its job.

Movies and TV shows played on my anxieties. I remember a *Twilight Zone* episode where a family ran to their fallout shelter when the air raid sirens sounded. The neighbors, who had no shelter, tried to get in. When the shelter owners refused, understandably, the neighbors grew increasingly agitated and finally broke down the door—just as the all-clear signal was given. I also remember seeing reruns of *Them*, a 1950s movie about giant ants created by nuclear tests, and *Godzilla*, a monster disturbed from his deep-ocean sleep by an atomic blast. *Fail Safe* depicted a nuclear war that began when a computer malfunction erroneously sent bombers to strike Russia, with no way to call them back. The Stanley Kubrick film, *Dr. Strangelove*, at least had a smattering of dark humor, featuring Peter Sellers as a crazed ex-Nazi and Slim Pickens, riding an H-bomb, waving his hat, cowboy-style, as it dropped from a plane.

The first atomic bomb used in warfare leveled Hiroshima in August 1945. *(National WWII Museum)*

Fears of a nuclear holocaust began in 1945, seven years before I was born. On August 6, the U.S. exploded Little Boy, a uranium bomb, 2,000 feet above Hiroshima, Japan. It destroyed everything within a one-mile radius, killing 70,000 people outright. Fat Man, a plutonium bomb, was dropped over Nagasaki three days later, killing another 40,000—over 110,000 Japanese citizens, mostly non-combatants, dead in three days.[1] But the nightmare wasn't over. Tens of thousands died in the next few months from radiation sickness, a grotesque scourge largely new to medical science. Harry Truman claimed the Japanese brought it on themselves, by not heeding his warning of a "rain of ruin from the air."[2]

Remarkably, in the immediate post-war years, the American public developed a morbid fascination with the "bomb." "Atomic Bomb Martinis" appeared in Las Vegas. Kids could save Kix cereal box-tops to get an "Atomic Bomb Ring" or buy the "Gilbert U-238 Atomic Energy Lab Kit" containing "low-level radiation sources." Beauty pageants crowned "Miss Atomic Bomb," with the winner wearing a white mushroom cloud dress, and in Las Vegas, "Dawn Parties" were held at casinos where guests watched early morning tests. Inspired by atomic tests at Bikini Atoll in the Pacific, a French designer came out with a new, skimpy bathing suit called the "bikini" to highlight its dangerous and explosive potential.[3]

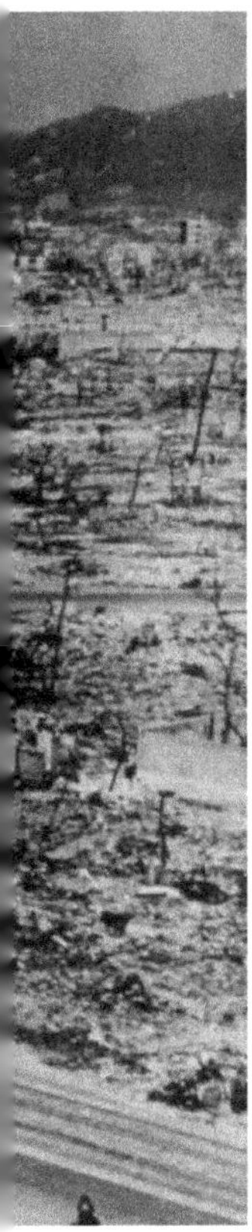

In 1949, after the Soviet Union tested its first atomic bomb, the competition for nuclear superiority was on, with both nations racing to stockpile more and bigger weapons. The U.S. military began testing new bombs at Yucca Flat, in the Nevada desert 65 miles northwest of Las Vegas. Early on, safety requirements for both observers and participants were lax. Troops stationed as close as one mile from the blast were instructed to charge into the "war zone" right after the explosion, bayonets fixed.[4] There was little in the way of protective gear, with soldiers simply instructed to shower and change clothes afterwards. A test in Nevada in 1953, later called the "Dirty Harry Shot," blanketed St. George, Utah, with as much radiation as nuclear workers were allowed in a year. Radios warned people to stay indoors and keep the windows closed.[5]

Almost all the early tests were atmospheric, with atomic devices exploded above ground. Seconds after detonation, megatons of vaporized materials were swept upward with the rapidly rising fireball to be caught and moved by the prevailing winds before settling to earth as fallout. Rainfall rinsed the atmosphere of radioactive particulates, often depositing them far downwind of the blast site. In April 1953, physicists at the Rensselaer Polytechnic Institute in New York noted a significant surge in radiation levels in rainfall thirty-six hours after tests in the Nevada desert, over two-thousand miles away.[6]

For a while, Missouri seemed to be in the bullseye for fallout. At least four bomb tests in the Nevada desert in 1955 and 1956 scattered radioactive fallout across Kansas and Missouri. After tests in 1955, Dr. W. D. Keller, a professor of geology at the University of Missouri, claimed his equipment detected gamma ray readings "twenty times the normal" from a rainwater puddle in Columbia.[7] After Soviet tests in the fall of 1961, atmospheric radiation detected in Jefferson City, Missouri, jumped from 25 micromicrocuries (mmc) to 450.[8] (The curie is a measure of the disintegration rate of radioactive atoms).

Fallout on crops could be eaten by humans, or on grass by cows. Of particular concern in fallout was strontium-90, a radioactive isotope chemically similar to calcium. Strontium-90 ingested in food or milk could replace calcium in bones and teeth, thus placing a long-lived radiation source in proximity to internal organs, potentially for years. This would greatly increase the cancer risk, especially for leukemia. In 1962, Missouri wheat had the second highest readings of strontium-90 in the U.S., with Iowa the highest.[9] Surveys by the United States Public Health Service in the early 1960s indicated elevated levels of strontium-90 in the St. Louis milk supply, the highest of ten cities tested.[10] Mothers worried about giving milk to their babies or children, and dairy farmers became concerned about their livelihoods.

The problem of strontium-90 in fallout was taken up by the St. Louis-based Committee for Nuclear Information, formed in 1958 to distribute scientific information about nuclear testing. Barry Commoner, who began teaching biology at Washington University in 1947, co-founded the Committee and became its most vocal and well-known member. He called atmospheric nuclear testing "the greatest single cause of environmental contamination on the planet." He acknowledged that scientists disagreed about the level of risk, but the "crucial point," to him, was that "the disagreements exist, for they reveal that we have risked these hazards before we knew what harm they might do."[11]

To help raise public awareness about the harmful effects of nuclear testing, the Committee in 1958 launched a "tooth survey," an idea suggested by a scientist in the journal *Nature* a few months earlier. Dr. Louise Reiss, a St. Louis internist, directed the survey. Parents of young children were asked to send in baby teeth for analysis, along with questionnaires about the child's age, the date the tooth was lost, the mother's residence during pregnancy, and whether breast-feeding or formula-feeding were used. The teeth were mailed with the forms to an office set up in St. Louis. The response was overwhelming, indicating a high level of concern. For several months, over fifty baby teeth arrived at the survey office every day. By the spring of 1960, 17,000 teeth had been received, and by the time the survey ended in 1968, over 200,000 had been collected.[12]

Dr. Reiss published the survey's preliminary findings in the journal *Science* in November 1961. The results largely confirmed the Committee's initial fears that strontium-90 levels in children's teeth had increased because of nuclear testing. The amounts doubled in the teeth of children born between 1951 and 1953, and tripled between 1953 and 1955. Levels gradually decreased after the cessation of atmospheric testing in 1963, when the three nuclear powers (the UK had joined the club by then) signed a limited test ban treaty, easing worries over fallout.[13] Commoner considered the treaty "the first victorious battle in the campaign to save the environment—and its human inhabitants—from the blind assaults of modern technology."[14]

Although atmospheric testing had stopped, subsurface testing went on, and the threat of nuclear war had not subsided. There were concerns about how to protect the populace, especially the children. In 1952, the New York Public School System issued 2.5 million dog tags to school children to help identify them (or their bodies) in the event of a nuclear attack.[15] Many people constructed fallout shelters, but most didn't. A Springfield Civil Defense official estimated in 1961 that "fewer than one percent of the families" in the city had them. In the event of a nuclear attack, he predicted, Springfield would be "reduced to an underground village of only a few hundred persons."[16]

But what would life be like for people who survived the initial blasts and subsequent fallout? Eminent scientists made dire predictions. Carl Sagan, using data from the Mariner 9 spacecraft showing a big drop in Martian surface temperatures during dust storms, predicted a "nuclear winter" in the aftermath of an attack.[17] Dust clouds would block sunlight for months, cooling the planet and stifling photosynthesis. "Numerous plant and animal species would disappear," Sagan suggested, and "vast numbers of surviving humans would starve to death." There was a real possibility of the "extinction of the human species."[18] It became apparent that surviving the initial blast and weeks of fallout would still mean an uphill struggle for human survival.

The arms race itself created jeopardy for many Americans. The number of people who died prematurely from the design, testing and construction of nuclear weapons is not definitively known, but the Department of Energy (DOE) employed hundreds of thousands of workers at over 350 secretive sites to assemble the eventual stockpile of 70,000 nuclear weapons. A Public Health Association report estimated that at least 33,500 workers at these facilities suffered higher than normal rates of dementia, miscarriage, nerve disorders, respiratory illnesses, and cancers.[19]

Sprawling Mallinckrodt Chemical Works in St. Louis, ca. 1955 *(State Historical Society)*

The Mallinckrodt Chemical Plant in St. Louis was one such facility. It took part in the Manhattan Project from 1942 to 1958, purifying uranium ore for nuclear weapons. Workers were exposed to a high radium-226 component in the uranium they processed. A study of 2,514 men who worked at Mallinckrodt from 1946 to 1966 found an increased risk of kidney cancers.[20] The Mallinckrodt site also became contaminated with radioactive waste, including uranium-235 and decay isotopes of radium, thorium, polonium, and thallium. Concerns first surfaced in 1946, when a reporter for the St. Louis *Post Dispatch* claimed the government and Mallinckrodt, when they reassured the public that the wastes were "not radioactive or otherwise dangerous," had told a "bald-faced lie."[21]

An estimated 121,000 tons of wastes produced at Mallinckrodt were transported off site and stored near the St. Louis Airport. Here, near the headwaters of Coldwater Creek, "thousands of barrels" of radioactive waste were stacked, "rusting and unattended for decades." Improper handling and transportation of wastes added to the problem. According to a reporter from the *Post-Dispatch*, "some truckers dumped their loads in empty fields to cut their workday short."[22] Hazardous wastes found their way into Coldwater Creek. Then, grading and housing construction in the 1950s and 1960s changed the course of the creek, which began eroding into storage areas and re-distributing waste materials, leading to the contamination of even more soil and water.

A local citizens' group, Coldwater Creek Facts, mapped and documented cases of cancer in the Coldwater Creek area, which appeared to be much higher than "normal." One member of the group who grew up near the creek, Debby Flondro, was diagnosed with ovarian cancer in 2012 at the age of 41. She and other residents canvassed the neighborhood and came up with a list of at least 57 cancer cases in people under 59 years of age who lived near Coldwater Creek. Many of them had already died. It was just "one funeral after another after another," Debby said; "they were 38, 39, 42 years old, and it just didn't make sense."[23]

Gail Vasterling, Director of the Missouri Department of Health, acknowledged in 2014 that "there may be something to the fears that have haunted the area's residents for decades."[24] She asked for federal assistance in evaluating the situation, and the U.S. Department of Health and Human Services launched a study in 2016. The report, released in 2019, confirmed that people living near Coldwater Creek from the 1960s to 1990s "may have" had an increased risk of developing bone cancer, lung cancer and leukemia.[25] Seventy-three years after the first complaints had surfaced, the government finally acknowledged that the health of nearby residents had most likely been affected by mishandled nuclear wastes.

Portion of the Mallinckrodt Pharmaceutical Plant in St. Louis *(Loring Bullard)*

Another radioactive waste problem in Missouri turned up at the Weldon Spring Ordinance Works, where explosives were manufactured for use in WWII. In 1940, the U.S. government acquired over 17,000 acres of land in St. Charles County, about twenty-five miles west of St. Louis. Almost 600 residents in the area were uprooted and forced to move, most of them from the little towns of Howell, Toonerville and Hamburg. By mid-1941, all the residents had been relocated. The Army paid for the construction of a plant to produce trinitrotoluene, or TNT, a powerful explosive. The plant, to be operated by the Atlas Powder Company, cost about $65 million. At its peak before it closed in 1945, it produced a million pounds of pure TNT per day.[26]

During the war years, the ordinance works had been an economic boon for the St. Louis region. With its closure, several ideas were put forward for reusing the huge site, which had an extensive water supply system and railroad network. The president of the St. Louis Chamber of Commerce, after the establishment of the United Nations in 1945, suggested the Weldon Spring site could serve as the "world capital."[27] But the military had other plans for it. In 1948, the House Armed Services Committee placed the Weldon Spring Ordinance Works in a "reserve pool to prevent possible sale as surplus property."[28]

Eight years later, 205 acres were transferred to the Atomic Energy Commission (AEC) to build a Uranium Feed Materials Plant for processing uranium ores into pure uranium trioxide, uranium metal and other compounds. The plant cost about $58 million and by the early 1960s, employed over 800 people.[29] Soon, there would be two kinds of toxic and extremely hazardous wastes at the site—nitroaromatic compounds in soil, sediment, and groundwater from former TNT manufacturing and radiological contamination from uranium processing waste, called raffinate, which was stored in four pits on the property.

Compounding the problem, some of the radiological wastes were also dumped into an old quarry located off-site, which had originally provided building stone for the TNT plant. From 1963 to 1969, the Army disposed of uranium and thorium in the old quarry, both in drums and uncontained. They also burned TNT production wastes and dumped contaminated rubble there. At the end of weapons material production at the Weldon Spring site, the quarry contained an estimated 96,000 cubic yards of radioactive and chemical wastes, as well as three million gallons of contaminated water.[30]

In 1966, uranium processing ceased at the Weldon Spring site, and the next year, the AEC turned the facility back over to the Army for use in producing defoliants—among them Agent Orange, a mixture of herbicides used widely in Vietnam. Luckily, that project was canceled before production began, or there could well have been a third type of hazardous waste to contend with—dioxin. The Army conducted limited remediation work at the Weldon Spring site in the early 1980s, but in 1985 gave full custody of the closed chemical plant

Rock "tomb" covers chemical and nuclear wastes near Weldon Spring. *(St. Louis Post Dispatch)*

to the Department of Energy (DOE).[31] The EPA placed both the radioactive waste pits and the quarry on the Superfund National Priorities list in 1987. The St. Louis *Post Dispatch* reported that the former quarry was "one of the three most hazardous" of the 99 sites added to the priorities list that year, which then totaled 802 sites nationwide.[32]

Remediation work under the DOE began in the late 1980s and continued until 2001, costing in total nearly one billion dollars. Wastes from TNT manufacturing were burned in an incinerator, including sixteen miles of buried wooden pipes containing explosives residue. 1.5 million cubic yards of radioactive wastes from the raffinate pits and quarry were moved into a specially engineered permanent storage facility, a "tomb," covering 41 acres. This massive rock-covered mound, 75 feet high, was constructed of compacted clay, synthetic liners, sand, gravel, and rock.[33]

The site is now open to the public. A long flight of stairs can be taken to an observation platform on top of the rock mound, affording outstanding views of the surrounding countryside, include a 150-acre restored prairie. An Interpretive Center was opened next door in 2002, where visitors can learn about the history of weapons production and the long road to final clean-up. In 2020, the USEPA awarded the site the National Federal Facility Excellence in Site Reuse Award.[34] About 8,400 acres of the original 17,000-acre Weldon Spring site is now managed by the Missouri Department of Conservation, with nearly twenty miles of hiking and biking trails.

My family moved from Kansas City to the quiet little town of Urich, Missouri in 1965. Just north of town, on a blacktop road in farming country, was a small military installation surrounded by a chain link fence, with corn growing on three sides. There was little inside the fence other than a concrete pad rising above the gravel and some vents or antennas sticking up. I knew what this was—a Minuteman Missile installation. Down in the ground, in a concrete-reinforced, steel-plated silo, beneath a 90-ton concrete and steel lid, stood a sleek, white, sixty-foot-high missile, six feet in diameter at the base and tipped with a nuclear warhead.

I drove by this missile site frequently. Normally, there would be no signs of life inside the chain link fence, but I knew people were there, below ground, sitting at their consoles in their hardened cocoon. Occasionally, I would see a dark blue, stretch-cab pickup truck parked inside the fence, and I knew there was something going on—maybe they were changing crews, or just getting lunch. The two-person crews made sure the missile was ready to fly within 55 seconds of receiving the launch order.[35] Only the president can give that order, and the crew members must each turn their keys simultaneously to initialize ignition. Key boxes are located twelve feet apart to prevent one person from launching the missile.

The U.S. military began installing these missiles in the spring of 1962. 150 missiles were placed in fourteen counties southeast of Kansas City, all controlled from the Whiteman Air Force Base in Knob Noster, Missouri. What I didn't know at the time was that of all the areas of the United States, central Missouri would suffer "the most devastating effect" of a Soviet strike against Minuteman missile bases, given their proximity to the large metropolitan areas of Kansas City and St. Louis. Other Minuteman missile installations might be targeted by the Soviets, but they were in areas of low population density in North Dakota, Wyoming and Montana.[36]

Minuteman missile launch facility underground control center *(National Park Service)*

We didn't have a fallout shelter at our house, but I remember my parents instructing us on where to go in the event of an attack. I knew that Urich was about fifty-miles from Kansas City, a likely target. I tried to imagine what the mushroom cloud would look like from that distance, and wondered if the flash would burn my eyes out—or whether the shock wave, like those shown in movies, would blow our house away. I imagined our family huddled somewhere underground, trying to weather the onslaught. Would we have enough food or water to survive? Where would we go to the bathroom? What would we see when we emerged? Any way it played out, the picture in my mind was bleak and depressing.

The nuclear arsenals of both the U.S. and the Soviet Union reached their peaks in the mid-1960s. After that, treaties began to reduce those numbers. The most sweeping of these agreements, the Strategic Arms Reduction Treaty (START) signed by George H. W. Bush and Mikhail Gorbachev in 1991, called for the elimination of 80% of the nuclear weapons then in existence, including the Minuteman II missiles in Missouri. These launch facilities were destroyed between December 1993 and May 1995.[37] After removing the missile and sensitive equipment, the Air Force turned the silo over to the Army Corps of Engineers for destruction. Silos were imploded, with the hole left open for 90 days to allow Soviet satellites time to verify the destruction. Then crews would then fill the hole and cover it with a concrete lid.

Family fallout shelters were highly recommended in the 1950s and 1960s *(The History Channel)*

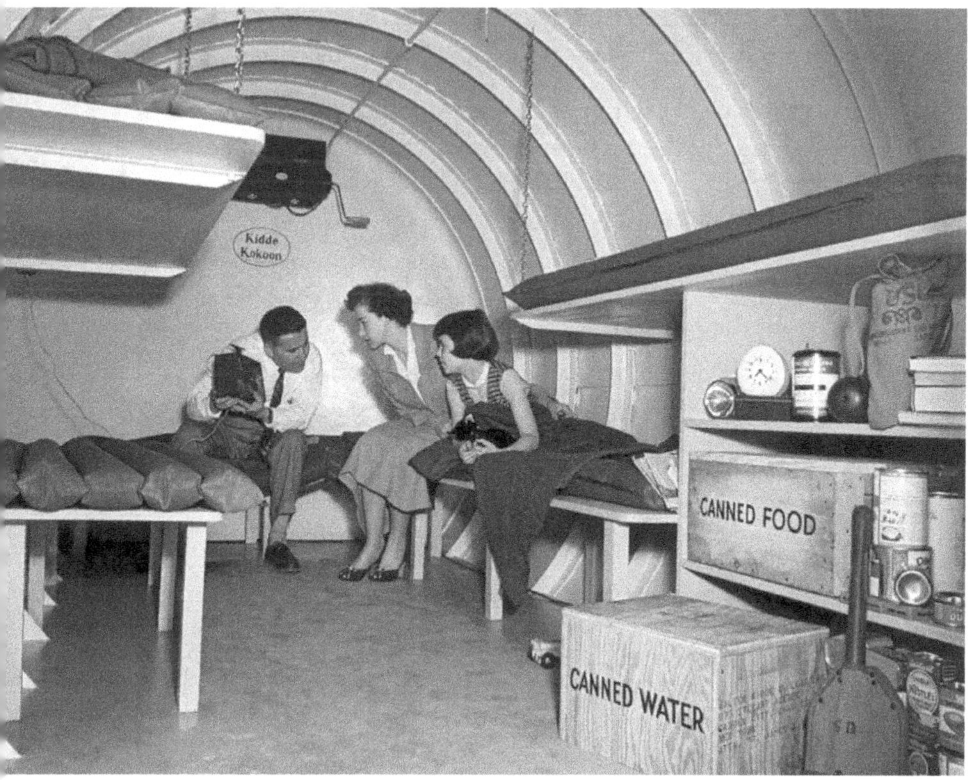

In 2023, I went back to Urich, my old hometown, to look for the site of the Minuteman II missile facility north of town. I couldn't find it, either on a drive-by or later, on Google Earth images. Apparently, it has been totally obliterated. However, I did find another former missile site in the area—about twenty miles south of Urich, near the little town of Taberville. The old chain link fence was still there, but rusty and vine-covered. Inside the fence were rows of big hay bales. Presumably, the site was being used for hay storage by a local farmer. It seemed a bit surreal—a little like a modern version of beating one's swords into plowshares. But way too many swords remain in other places, the use of any one of which would mean a human and environmental disaster of epic proportions.

It has been argued that the fear of thermonuclear war helped to spur the environmental movement. Concern about a possible nuclear exchange added to society's anxiety levels, reinforcing the terrifying truth that our thin and fragile biosphere could be devastated in an instant. In 1961, Margaret Mead observed that "growing up in the shadow of the bomb" made young people more likely to understand the environmental crisis. "They have never known a time when war didn't threaten annihilation."[38] Having grown up squarely in the shadow of the bomb, I can attest to the fact that it increases a person's anxiety level. For most of my life, I have felt an underlying nervousness about all those weapons, locked and loaded, poised to destroy all that I know and love. That was the substance of nightmares I had at the age of eleven, searching for shelter in my fitful sleep as the rain of ruin came down.

As I grew older, I came to realize the horror of nuclear war called into question the sanity of the human race. There were millions of people who agreed that such an event must never happen. They called on their governments to stop the madness, end the arms race, sign treaties with other nations to eliminate nuclear weapons. But that wasn't all. There was the matter of the seemingly endless war in southeast Asia, where thousands of young Americans were dying, and the reckless endangerment of people at home by pesticides and water and air pollution. It was all reaching an explosive climax when I began my last year in high school, and soon thereafter a burning river and spewing oil well lit the fuse.

Abandoned Minuteman missile launch facility in Henry County *(Jim Guldin)*

SOURCES

1 Allan M. Winkler, *Life Under a Cloud: American Anxiety About the Atom*, New York, 1993, p. 22

2 Ibid., p. 24

3 Ibid., p. 28

4 Ibid., p. 92

5 Michael Egan, *Barry Commoner and the Science of Survival*, Massachusetts Institute of Technology, 2007, p. 51

6 Richard L. Miller, *Under the Cloud: The Decades of Nuclear Testing*, 1991, p. 170

7 *The Daily Standard*, May 11, 1955, p. 1

8 *St. Louis Post-Dispatch*, September 23, 1961, p. 74

9 *Under the Cloud,* p. 354

10 *Barry Commoner and the Science of Survival,* p. 67

11 Ibid., p. 26

12 Ibid., p. 70-71

13 *St. Louis Post-Dispatch*, June 8, 1969, p. 1

14 *Barry Commoner and the Science of Survival*, p. 47

15 Ibid., p. 115

16 *Springfield Leader and Press*, September 10, 1961, p. 43

17 *Life Under a Cloud*, p. 200

18 Ibid., p. 201

19 *The Kansas City Star*, December 11, 2015, p. 1.

20 Robert Alvarez, "A Review of the Health and Mortality Experiences of U.S. Dept. of Energy Workers," American Public Health Association, 2000

21 *St. Louis Post Dispatch*, August 11, 2017, p. 13

22 Ibid., p. 13

23 *St. Louis Post Dispatch*, February 29, 2012, p. 5

24 *St. Louis Post Dispatch*, October 3, 2014, p. 10

25 *St. Louis Post Dispatch*, May 1, 2019, p. 3

26 *St. Louis Post Dispatch*, December 1, 1993, p. 21

27 *St. Louis Post Dispatch*, November 6, 1945, p. 3

28 *St. Louis Star and Times*, July 7, 1948, p. 54

29 *St. Louis Post Dispatch*, February 22, 1962, p. 13

30 *St. Louis Post Dispatch*, July 22, 1987, p. 119

31 U.S. Dept. of Energy, "Weldon Spring Site, Missouri" fact sheet, 2020

32 *St. Louis Post Dispatch*, July 22, 1987, p. 119

33 Website, dnr.mo.gov/waste-recycling-sites-regulated-facilities/federal/Weldon-spring-site

34 U.S. Dept. of Energy, "Weldon Spring Site, Missouri" fact sheet, 2020

35 *Urich Herald-Montrose Tidings*, May 4, 1962, p. 1

36 *The Kansas City Times*, January 16, 1975, p. 1

37 Website, nps.gov, "Deactivation of the Minuteman II Weapons System" (U.S. National Park Service)

38 Adam Rome, *The Genius of Earth Day*, New York, 2013, p. 38

CHAPTER 3:
THE PROMISE OF EARTH DAY

"The ultimate test of a man's conscience may be his willingness to sacrifice something today for future generations whose words of thanks will not be heard."

GAYLORD NELSON

To environmentalists, Earth Day is a red-letter day, marking a major milestone—a critical turning point in history. To many others, the day is barely noticed, just another minor "holiday" to ignore. Even for many of the believers, what started out as a definitive moment, an uprising, has become overly institutionalized, now bringing routine and unremarkable obligations—clean-ups and work projects, outreach sessions with schools. For many, the original context and significance have been lost. What started as a show of force—unbridled, emotionally-driven activism, with millions marching in the streets—has been transformed into a watered-down, light-hearted event, more a subdued celebration of past successes than a call to action.

Even the idea of Earth Day is unsettling to some environmentalists. To them, thinking about nature for only one day a year seems overly superficial, a mere passing nod to Mother Earth. We might feel a tiny quirk of guilt for not doing more, or be reminded to resume recycling, or cut down on water use, or buy an electric lawnmower, but we soon return to our old ways; our conspicuous consumption. Still, Earth Day is, and always has been, at least partly about touching the stone— a reminder that we can, and must, keep the world healthy and livable for future generations—those who will inherit our fragile planet, for better or worse.

By almost any measure, the first Earth Day in 1970 was a huge success. An estimated 20 million Americans turned out, nearly 10% of the population. Over 2,000 colleges and 10,000 K-12 schools took part.[1] Denis Hayes, the primary coordinator of Earth Day, called it the "largest, cleanest and most peaceful demonstration in American history."[2] Earth Day events ran the gamut, from street festivals, clean-ups, and neighborhood fix-ups to seminars at schools and universities. In New York City, major thoroughfares were closed to accommodate an estimated crowd of one million. Fifth Avenue, closed to vehicular traffic, became a "park" where Pete Seeger and Leonard Bernstein performed.

For its organizers, Earth Day had the desired effect, creating a powerful spotlight on the need for change. Politicians at federal, state, and local levels took note. Congress, sensing the magnitude of the uprising, was galvanized to act. Almost every environmental measure before federal lawmakers in the early 1970s gained an amazing level of bipartisan support. In 1970 alone, in rapid succession, Congress passed the Water Quality Improvement Act, the Resource Recovery Act, and Clean Air Act amendments. Never had so many environmental laws been written and passed within a single year.

The successes of the movement changed the culture of environmentalism. Environmental attorneys proliferated to interpret and enforce the new laws. Environmental groups hired lobbyists to roam the halls of Congress. Many industries got on board with "green" campaigns. Several "Ecology Centers" were opened, including one in St. Louis. Environmental education curricula proliferated at colleges and universities. The media consistently featured environmental stories and increased staffing to do so. Groups like Audubon, Sierra Club and Natural Resources Defense Council enrolled over 300,000 new members in the first year after Earth Day.[3]

Five months after the first Earth Day, I started to school at Central Missouri State College in Warrensburg, majoring in biology. One of my first classes was taught by Dr. Oz Hawksley. With his square black glasses and silver hair combed back, Hawksley looked the part of the nutty professor. But he was wiry and tough, and earlier in life had been a whitewater boatman on western rivers. I will never forget the first time I saw him in class. He wore a blazer, and as he pointed to the blackboard, his jacket would open slightly, exposing on his chest a large white button with three black letters—ZPG. I had no idea what it meant, and no impulse to ask.

In the ensuing months, I would find out more about the button and the man behind it. Some of it happened in the classroom, as Hawksley paused his lectures on ice-age bears or arctic terns to talk about the ecological effects of dams on rivers. He would suddenly launch into a discussion about the human population explosion. There were just too many people competing for finite resources, he said. He advocated for birth control and the organization formed to promote it—ZPG, or Zero Population Growth. And he urged us to take part in the second Earth Day celebration that would be held on campus the following spring.

I got to know Oz better when I joined the school's Outing Club. As the faculty sponsor, he led us on floating and backpacking trips and taught us outdoor camping and cooking skills. He instilled a love of rivers and the outdoors in me, but more importantly, he awakened my environmental consciousness. No matter what career we pursued, he told us, we needed to help save the planet from greedy corporations and thoughtless polluters. A good way to do that, he said, was to get active in the movement that was just then getting underway. By the second Earth Day on April 22, 1971, I had joined that movement, at least philosophically.

By then, major disasters had pushed the environment front and center. In January 1969, eight days after Richard Nixon took office, an oil rig in the channel off Santa Barbara, California ruptured, spilling crude oil over thirty-five square miles of ocean. Over 3,000 birds perished. Images on TV of shivering, oil-soaked birds alarmed and infuriated the public. Nixon's White House Advisor, John Whitaker, said the spill was "comparable to tossing a match into a gasoline tank; it exploded into the environmental revolution, and the press fanned the flames to keep it burning."[4] Nixon supported offshore drilling, but saw the disaster as an opportunity to make points with the growing ranks of environmentalists.

The first Earth Day in New York, 1970, attracted a huge crowd *(EarthDay.org)*

Thousands marched in the streets on the first Earth Day, April 1970 *(NYC Dept of Records)*

That summer, another galvanizing event occurred when the Cuyahoga River in Ohio caught fire. Winding through the city of Cleveland, the Cuyahoga is lined with steel mills and other industries that dumped massive volumes of waste into the river, making it so thick with oily scum that one reporter suggested the water "oozes rather than flows."[5] Lake Erie, which receives the Cuyahoga's flow, was said to be "dying" from all the industrial pollution and algae-stimulating nutrients from inadequately treated sewage. Only three of the 62 public beaches on the lake were open for swimming that summer. The stage had been set for a high-profile disaster. It happened when a spark from a train crossing the Cuyahoga ignited the floating scum of oily waste below.

It was a small fire compared to some of those in the past, but the images of a river literally burning further inflamed a public already disgusted with the oil spill just six months earlier. *TIME* magazine made the burning Cuyahoga the "poster child" of polluted waters. John Ehrlichman, Nixon's Assistant for Domestic Affairs, told the president the disaster afforded him a unique opportunity to push an environmental agenda, the "moonshot of the moment."[6] A savvy politician, Nixon recognized the growing political power of the environmental movement. He decided to get ahead of Democrats and take credit for new environmental legislation. On New Years Day 1970, he signed the National Environmental Policy Act (NEPA), now one of the nation's most important environmental laws.

The oil spill in California and the fire on the Cuyahoga set other plans into motion. In the fall of 1969, having inspected lingering damages from the Santa Barbara spill, Senator Gaylord Nelson was flying to Seattle to address the Washington Environmental Council. En route, he read in a newspaper about the Vietnam War "teach-ins" being held on college campuses. In contrast with many of the militant protests then occurring, the calm, rational nature of teach-ins impressed local elected leaders. Nelson decided right then that he would spearhead a national teach-in for the environment. "I am convinced that the same concern the youth of this nation took in changing this nation's priorities on the war in Vietnam and civil rights can be shown for the problems of the environment."[7]

It turned out to be a stroke of genius. Nelson's plan would leverage the youthful energy and enthusiasm that had so effectively powered the civil rights and anti-war movements. Before this time, students and young people in general had been largely ignored by politicians, especially since the voting age was 21 at the time. Nelson hired a small staff, opened an office, raised funds, and created a non-profit organization, all in three months. The teach-in would be decidedly grassroots, with no formal direction from the top. Nelson and his staff provided only the date of the event, national publicity, and some suggestions on what topics could be discussed.

To maximize participation, the date chosen for the teach-in was April 22, after the college spring break and before final exams. Denis Hayes, a young activist from Washington state, was hired as the event coordinator. Hayes took his cues from recent anti-war protests, announcing at his first press conference in January 1970 that "we cannot save the environment as long as the war goes on." With Nelson's blessing, organizers soon changed the name of the event to Earth Day, and to publicize it, took out a full-page ad in *The New York Times*. Hayes' lofty goal was to involve "an entire society in a rethinking of many of its basic assumptions."[8]

Compared to the impressive scale of events in New York and other big cities, Missouri's Earth Day offerings were relatively low-key. Nevertheless, most of the communities and schools in the state took part. St. Elizabeth High School in Miller County "plunged" into the fight against litter pollution with "practical action," devoting the day to picking up trash along thirty miles of roads.[9] Lincoln University in Jefferson City held lectures on "Logical Techniques in Environmental Problems" by the Philosophy Department and "Planned Parenthood" by the Nursing Department.[10] At a teach-in at the University of Missouri-Kansas City, Attorney General John Danforth, addressing a crowd of 350, called for the creation of "nationwide standards on pollution" and the enforcement of "strict time limits" to comply with laws.[11]

Earth Day became an annual event, a permanent date on the calendars of most cities and schools and every environmental organization. The twentieth Earth Day, in April 1990, was much bigger than any since 1970. The effort was again led by Denis Hayes, coordinator of the first Earth Day, who in 1990 had a well-paid staff and a budget of $3 million.[12] The celebration became international, attracting an estimated 200 million people in 138 countries. Almost 200,000 people gathered in Washington D.C., where Gaylord Nelson, now 74, said the goal of Earth Day was to "force the political establishment to catch up to the people."[13]

As in 1970, Earth Day events in Missouri in 1990 were widespread, but this time were more organization-driven. All of Missouri's major cities, along with 3,000 communities nationwide, participated in some way, with parades, concerts, festivals, and work projects. In Kansas City, "eco-troubadour" Stan Slaughter sang and picked up trash, just like he had done twenty years earlier. A parade wound from Volker Park to the Country Club Plaza, featuring dragonflies with twenty-foot wingspans, gorillas, ladybugs and "South American caterpillars."[14] The city of Ava in southern Missouri had a pet contest and solar composting demonstrations.[15] In St. Louis, there was a "Rock for Life" concert, a "Global Forest Festival," and an "All Species Parade."[16]

There were plenty of speeches, of course. Missouri Governor John Ashcroft issued a challenge to the state's communities to plant a million trees. The "so-called greenhouse effect, or global warming," he said, is a "result of fewer trees."[17] Jerry Presley, Director of the Missouri Department of Conservation, announced that his agency would offer $100 "incentive coupons" to 950 communities willing to buy "street-side trees."[18] G. Tracy Mehan, Director of the Missouri Department of Natural Resources (DNR), described an uneven struggle to protect the state's resources. Between 1970 and 1990, he said, the state had "seen its wetlands shrink, its skies clear, and the health of its major rivers decline." Despite its environmental problems, Missouri's legislators were not in a regulatory mood in 1990. DNR's budget had been repeatedly cut, from about 4% of general revenue in 1970 to 2% in 1990.[19]

To many Earth Day participants, the event lost much of its luster—its fizz—over time. As it became less grassroots driven, it seemed less exciting. David Brower, Director of Friends of the Earth, said Earth Day had "started out as an orgasm, and then they couldn't quite repeat that kind of success."[20] On the thirtieth anniversary in 2000, a writer for *The Kansas City Star* suggested Earth Day had become "something like the green version of Secretary's Day. Everyone is for it. Everyone feels good about it. But that broad support is of questionable depth."[21] Earth Day had become like New Year's Eve, another writer sneered; "we think about it once a year, make wild promises we will never keep, and then go home disappointed."[22]

To many environmentalists, once-per-year participation in an Earth Day event represented a superficial notion of what it meant to "save the planet." Taking part made people feel good about their environmental intentions, but represented just one small step toward true and lasting stewardship. The long-term commitment often wasn't there. And part-time environmentalists had short attention spans. The public overwhelmingly supported pollution control laws in 1970, but polls taken after the oil embargo in 1973 showed Americans were already more worried about an energy shortage than about pollution.[23]

Many people criticized Earth Day for failing to focus on the most critical issues. By attempting to be inclusive, the event seemed chaotic, scatter-brained. Was it about pollution? Public health? Wildlife protection? Social justice? St. Louis environmentalist Barry Commoner said this lack of coherence created confusion around the event. Earth Day, he observed, "mirrored personal convictions more than objective knowledge," convincing him of the "urgency of a deeper public understanding of the origins of the environmental crisis and its possible cures."[24] Decades after the first Earth Day, he lamented the "failure of a social movement" that in 1970 seemed "poised to incite significant change."[25]

On the other hand, by acting on pollution first, the movement avoided the thorny, controversial issues of population control and environmental justice. With a focus on clean water and air and food safety, Earth Day environmentalism appealed to the middle-class "conservative, silent majority," who largely felt alienated by the social activism of the 1960s.[26] But social activism was what Commoner saw as the greatest need. "Ecological crusades" against overconsumption, he said, had made environmentalism "irrelevant" to advocates of social justice, especially when polluting industries were still locating in poor neighborhoods of color.[27]

Cuyahoga Fire: In 1969, the burning Cuyahoga River in Ohio brought nationwide attention to water pollution *(The Allegheny Front)*

The movement was criticized for not making stronger alliances with minority groups. There was a prevailing view that it was supported and led by young, white, rich kids. A CBS program on the first Earth Day, "A Question of Survival," looked at the "skepticism of young black radicals" toward the "ecology movement."[28] To demonstrate their disgust with excessive consumption, students at San Jose State College in California buried a new, $2,500 car on the campus, but the event was picketed by black students who thought the money would've been better spent helping poor people in the ghettos.[29] The April 1970 issue of *Ebony* magazine, featuring Martin Luther King Jr. on the cover, did not even mention Earth Day.[30]

Over time, the priorities of Earth Day changed, reflecting not only progress on some pollution fronts, but also an awareness of new challenges. At a 20th Earth Day anniversary event, Gaylord Nelson said the battlefield had shifted from water and air pollution to land, "where landfills overflow, hazardous wastes leach from abandoned dump sites, and nuclear wastes accumulate with no place to put them."[31] While scientists and environmentalists were aware of these new threats, the public, for the most part, was not. Participants at later Earth Day events were still worried most about air and water pollution—problems that had largely been addressed—rather than timely issues like global warming and urban sprawl.[32]

For many who took part in Earth Day every year, the most important element was environmental education. In 1970, President Nixon promoted environmental literacy, the teaching of environmental concepts "at every point in the education process."[33] Every year, schools were the single largest group of Earth Day participants, with millions of kids taking part in classwork, games and hands-on activities. Gaylord Nelson thought the best results from Earth Day came from teaching grade school kids. "If they become interested, the impact extends all their lives."[34] But to some, the focus on youth appeared to make Earth Day into fluff. A writer for the Columbia *Daily Tribune* suggested in 2007 that Earth Day had become about "human interest photos" of kids "launching a flotilla of buoyant recycled materials across a murky pond," which the next day were buried in the local landfill with a "half-ton of other Earth Day celebratory debris."[35]

Despite such views, those of us who participated in Earth Day events noticed that most young kids were interested and enthused. It made us feel better about the future. But now, fifty years out, we wonder why the environmental ethic hasn't become more pervasive. Did we really instill a life-long interest in protecting the environment, or did we just provide a fun day for kids and a relaxing diversion for their teachers? Where are the voices that we thought by now would be speaking up on the Earth's behalf? Did environmentalists lose the battle for the hearts and minds of today's adult Americans?

Earth Day 2019 celebration in Springfield included creating a likeness of the Earth using colorful umbrellas *(Public Information Office, City of Springfield)*

In 2019, I helped organize the 49th Earth Day event in Springfield, held downtown on the Square. It was meant to be a "warm up" for the blockbuster 50th anniversary of Earth Day to be held the next year, in 2020. But COVID hit, and the 50th event had to be scrapped. I understood, but still felt a little cheated, having participated in some way in almost every Earth Day since 1971. On April 22, 2020, I decided to go floating instead. I loaded up my canoe and headed to a nearby Ozark stream. I took along a trash bag, as I always do.

I can't remember how much trash I picked up that day, or what it was. But I do remember thinking that the simple action of lifting a beer can or plastic bottle from a beautiful stretch of river provides a moment to reflect on why Earth Day is so important. I thought about what Gaylord Nelson said at the 1990 Earth Day event in Washington D.C.: "I don't want to come back here twenty years from now and have to tell your sons and daughters that you did not do your duty."[36] I have two daughters, and now grandkids, and I don't ever want them to see burning rivers or beaches begrimed with oil sludge or animals going extinct from chemical poisoning or habitat destruction.

Every Earth Day since 1970 has been a mixed bag—there are always some successes to celebrate, some failures to report. We celebrate the fact that, overall, we have made great strides in improving the environment. We owe a debt of gratitude to our predecessors for changing their ways, even as we recognize their failings. In large measure, Earth Day is still about hope. When I begin to lose hope that we will ever solve today's seemingly insurmountable environmental crises, I think about how bad the air was in St. Louis and Kansas City in the 1930s and 1940s, or how nasty and polluted many of our rivers and lakes were in the 1950s and 1960s.

To me, an important lesson from the first Earth Day, and the 1970s in general, is this: good laws are necessary and they work. In 1969, there were no oil executives shouting, "we need more controls on offshore drilling;" no industrial magnates saying "we really need to stop dumping our wastes into the Cuyahoga." The current "roll-back regulations" mood of many in Congress angers me. Demanding a universal roll-back ignores the fact that these rules came about to solve real problems and prevent new ones. To flippantly suggest doing away with them is a slap in the face to our Congressional predecessors who finally took a stand against environmental degradation.

The first Earth Day was a long time ago. Fewer and fewer people remember what it was like back then. In 1970, nearly every state had polluted rivers and cities with foul air and wild places being overrun with development or heavy-handed resource extraction. Widespread degradation created a groundswell of citizens demanding change, a tidal wave of public opinion that was hard for politicians, both Republican and Democrat, to ignore. By Earth Day 1990, in contrast, things had gotten a lot better. Environmental problems still existed, of course, but there were no rivers catching on fire or air so polluted it was dangerous to breathe. It became easy to forget the darkest days of environmental abuse.

Today, however, new and even bigger threats have emerged. They affect us all, whether we realize it or not—the biodiversity crisis, where we risk losing critical ecosystem services such as crop pollination and firewalls against the spread of disease; and the climate crisis, which threatens young and old, rich and poor alike, from a California millionaire living in the fire-prone foothills to the lowest beach shanty in the path of a hurricane. But powerful agents are at work to reassure us there is little reason for worry—telling us that the environmentalists are once again crying wolf. With the problems of the past now only dim memories, a new awakening is desperately needed—and vigorous action to confront today's deepening crises.

SOURCES

1 Adam Rome, *The Genius of Earth Day,* New York, 2018, p. 209

2 J. Brooks Flippen, *Nixon and the Environment*, Albuquerque, 2000, p. 14

3 *Nixon and the Environment*, p. 99

4 Ibid., p. 566

5 Ibid., p. 574

6 Ibid., p. 574

7 *The Genius of Earth Day,* p. 57

8 Ibid., p. 87

9 *The Miller County Autogram-Sentinel,* April 30, 1970, p. 2

10 *Jefferson City Post-Tribune,* April 22, 1970, p. 1

11 *Kansas City Times,* April 22, 1970, p. 3

12 Douglas Brinkley, *Silent Spring Revolution*, New York, 2022, p. 276

13 *St. Louis Post Dispatch,* April 22, 1990, p. 25

14 *The Kansas City Star,* April 23, 1999, p. 11

15 *The Douglas County Herald,* April 19, 1990, p. 1

16 *St. Louis Post Dispatch,* April 1, 1990, p. 17

17 *St. Louis Post Dispatch,* March 28, 1990, p. 192

18 Ibid., p. 192

19 *Springfield News Leader,* April 23, 1990, p. 56

20 *Nixon and the Environment*, p. 167

21 *The Kansas City Star,* April 23, 2000, p. 8

22 Ibid., p. 41

23 *Nixon and the Environment*, p. 207

24 Barry Commoner, *The Closing Circle*, New York, 1972, p. 10

25 Ibid., p. 139

26 Ibid., p. 105

27 Ibid., p. 177

28 *St. Louis Post Dispatch*, April 21, 1970, p. 6D

29 *The Closing Circle*, p. 207

30 *Barry Commoner and the Science of Survival*, p. 175

31 *St. Louis Post Dispatch*, April 22, 1990, p. 25

32 *The Kansas City Star,* April 23, 2000, p. 8

33 *The Genius of Earth Day*, p. 230

34 *Springfield News-Leader*, April 21, 2000, p. 22

35 *Columbia Daily Tribune*, April 22, 2007, p. 41

36 *Springfield News-Leader*, April 23, 1990, p. 56

CHAPTER 4:
OLD KING COAL

"Earth provides enough to satisfy every man's needs, but not every man's greed."

MAHATMA GANDHI

My high school years were spent in Henry County, Missouri—a region of black soil, cornfields, coon dogs and coal. Most of the coal had already been dug up before I moved there, but there were unmistakable signs of its passing, especially a few miles south of Urich, my hometown, where heaps of gray-brown earth and sandstone rocks lay in hummocky rows like gargantuan mole runs. These piles were the spoils, the leftover materials, the "overburden" in the parlance of miners. Big shovels had been there a decade earlier, some of them as big as office buildings. Peabody Coal, by the 1960s the only coal mining company left in Henry County, operated some of the largest shovels in the world, such as "Big Hog," the Bucyrus monster commissioned in 1962. At 9,000 tons, it was at the time the heaviest land machine ever built; twenty stories tall, chewing out 160 tons of rocks per bite in a dipper holding 115 cubic yards.[1]

Coal had been mined in Henry County long before the big shovels showed up. Right after the Civil War, coal was mined commercially near the little town of Lewis, northeast of Clinton. This coal was taken out by pick and shovel from a shaft forty feet below the surface. Like others nearby, this mine was eventually bought out by Peabody. By then, there was no more digging underground. Instead, the coal was all excavated from surface mines. To get to the coal, the overburden, the soil and rock "waste" over the coal seam, was stripped off and dumped into piles at the edge of the pit. The piles became long ridges as the shovel worked its way along a linear path, gouging out a long, narrow trench. When the coal had been removed from one section, another was opened nearby, leaving clusters of parallel scars on the land.

The "strip pits" remained open, and over time filled with water. Decades later, they became cool summertime swimming holes for sweaty teenagers like me. The water was clear, but also acidic. I remember stinging eyes after swimming and iron-stained, oil-sheened water around cattails in the shallows. Near one of the pits was an abandoned shovel, smaller than Big Hog, but still immense. My friends and I would climb rusty ladders to reach the main cab, far above the ground, for views of coal mining scars and cornfields in all directions. We gave little thought to the environmental degradation represented by that huge machine and the surrounding, water-filled strip pits—not just the ripping up of the earth, but the burning of coal and its effects on human and environmental health.

Missouri was never a big coal producer, but it was the first state west of the Mississippi to mine coal commercially. Missouri's bituminous or "soft" coal was considered inferior to "hard" anthracite coal, mined mostly in Pennsylvania. Anthracite burned hotter and with less sulfur-polluting smoke, but it was more expensive. Missourians burned mostly bituminous coal, much of it from Illinois, and by the turn of the 20th century, coal smoke had seriously polluted the air in the state's larger cities, with thousands of individual homes and businesses burning coal in stoves, furnaces, and boilers.

Bevier Coal Company's massive shovel, 1952 *(State Historical Society)*

St. Louis passed the first ordinance on nuisance smoke shortly after the Civil War, the first major city in the U.S. to do so. But a rapidly increasing population added more and more sources of coal smoke, overwhelming attempts at control. The city passed a series of more stringent ordinances in the early twentieth century, but none were very effective. Enforcement was usually lax, with authorities disinclined to punish offenders. Poor people could not afford more expensive "smokeless" coal and businesses complained about the costs of mechanical smoke-reducing devices.[2] To help achieve compliance, private citizens, especially women's organizations, got involved in enforcing the ordinances. In 1907, the Women's Organization for Smoke Abatement published lists of non-compliant businesses in the newspapers.[3]

Smoky skies over St. Louis, 1939
(St. Louis Post Dispatch)

By the late 1930s, smoky conditions in St. Louis had grown intolerable. On November 28, 1939, later referred to as "Black Tuesday," the worst smoke and fog blackout "of the season" descended on the city. Heavy smoke turned day into night, with photographs showing vehicles with lights on at midday and buildings barely visible. Smoke seeped into the city hospital, "causing discomfiture to patients, interns and nurses alike." The Trudeau Society, an organization of pulmonary specialists, said this and other "smoke palls" had caused the lungs of St. Louisans to "more nearly resemble lumps of coal than the clean pink organs with which they were born."[4]

Black Tuesday marked a turning point. An active citizen-based committee had already made a series of smoke abatement recommendations to the St. Louis Board of Alderman. In April 1940, under pressure from over a hundred civic groups, the board approved a new ordinance based on the committee's recommendations. This law had the requisite teeth. Residents were required to either buy approved combustion devices or use smokeless coal. The market for cleaner burning coal expanded after this, providing a cheaper alternative for those who couldn't afford cleaner-burning mechanical devices.

Smoke battles were also being fought in Kansas City. A representative of the coal industry in 1941 stated his opinion that the city's problems could be solved by enforcing the existing ordinance, with slight changes in provisions, "without the drastic methods taken by St. Louis."[5] But things did not improve. In October 1945, the city "choked and coughed through a heavy blanket of smoke and fog." The Weather Bureau rated visibility at "zero" and the newspaper reported that "not a plane landed or took off at the Municipal Air Terminal." Kansas City's Chief Boiler and Smoke Abatement Inspector expressed his hope that a "new smoke ordinance resembling that of St. Louis" would pass in Kansas City.[6]

Kansas City was among 84 cities that sent representatives to St. Louis in 1941 to learn about its smoke abatement rules. Within ten years, most of these cities had established their own ordinances. Stronger ordinances helped, but the biggest changes would occur when industries and homeowners switched from coal to more efficient, less polluting fuels. In 1949, a large natural gas pipeline was completed into St. Louis from well fields in the southern U.S. By the early 1950s, only 30% of homes in St. Louis still used coal-fired heating, while almost 60% used natural gas or oil.[7]

When I moved with my family to Henry County in 1965, the only active coal mine was near the town of Montrose. Nearly two million tons per year were still being mined there to supply a coal-fired power plant, a 500-megawatt station producing electricity to serve the Kansas City region. Surface mining to supply the power plant began in 1956. It was dirty, bituminous coal, but it was cheap. A Kansas City Power and Light spokesman said it cost "half as much as Wyoming coal."[8] Since the mines were right at the power plant, transportation costs were minimal. By the late 1980s, there were 2,400 acres of abandoned surface-mined lands near Montrose. This included 1,200 acres of barren, acidic soil and 100 lakes in the abandoned pits. More than half of the lakes were very acidic, with a pH of 4.0 or less.[9]

This problem wasn't unique to Henry County, of course, since coal was mined at many sites in western and northern Missouri. Until 1971, coal mining was virtually unregulated in the state. Most of the old mined over lands remained scarred and pitted, nearly unusable. These lands were too rough to support farming, and resource agencies noted fish kills and other problems in creeks receiving acidic drainage from the mines. Missouri's DNR reported "severe water problems" at mined lands in the northern and southwestern parts of the state, where "stream life has been decimated by acid runoff from abandoned coal mines."[10] By the late 1960s, the problem was receiving attention nationwide, with many calling for the federal government to step in and regulate the industry.

In 1977, President Jimmy Carter did just that when he signed the Surface Mining Control and Reclamation Act, which directed mining companies to submit a reclamation plan prior to mining. After extracting the coal, they were required to return land to the approximate contours that were present before mining, stockpile topsoil for replacement, and re-vegetate barren areas. Missouri is eligible for half the revenues from the 35 cents-per-ton fee levied against mining companies for reclamation work. In a typical reclamation project, spoil piles are graded to a gentle slope, treated with agricultural lime to raise the pH, and covered with suitable soil to promote successful revegetation. There were about 66,000 acres of abandoned strip mines in the state when the law went into effect in 1980. As of 2024, about 5,000 acres had been reclaimed. Most of the remaining acreage will be allowed to "regenerate naturally."[11]

Like mining, coal burning would eventually face stricter rules. In 1963, Congress passed the Clean Air Act, the first federal law superseding state laws and taking direct action to control pollution. Major amendments in 1970, 1977 and 1990 further strengthened federal authority and addressed pollution from stationary sources like coal-fired power plants. Clean air bills had strong political support. The 1970 Clean Air Act passed the House by a vote of 374 to 1 and the Senate by 74 to 0. Missouri's two Republican senators, Christopher Bond and John Danforth, both voted for the 1990 Clean Air Act in a lopsided vote of 89 to 10.

The Clean Air Act worked as intended. Between 1980 and 2000, the U.S. cut its emissions of sulfur dioxide by 30 percent, much of it from coal-fired power plants. Missouri's relatively dirty coal fell out of favor after passage of the act because it forced power plants to install expensive pollution control equipment. Where Missouri coal contains about four pounds of sulfur per million BTUs of energy produced, Wyoming coal has less than one-half pound.[12] Most power plants in Missouri now burn Wyoming coal.

Downtown St. Louis, November 28, 1939, called "Black Tuesday" *(St. Louis Post Dispatch)*

Despite burning cleaner coal and installing pollution control equipment, Missouri still produces a disproportionate share of the remaining sulfur dioxide pollution in the U.S. Our state has the highest emissions of the four states in EPA region 7—Missouri, Kansas, Iowa and Nebraska. Missouri power plants released 91,000 tons of sulfur dioxide in 2020. Ameren's Labadie Plant produced almost 40,000 tons of it, the second highest of any plant in the country.[13] However, EPA says these plants now comply with air emission limits of the Clean Air Act.

A bulky and potentially harmful residue from the burning of coal is the leftover ash. It is primarily made up of calcium, silica, aluminum, and iron, but may also contain minute quantities of toxic constituents, including chromium, mercury, lithium, selenium, and arsenic. The scale of the problem is massive. Coal plants in the U.S. annually produce 70 million tons of ash. Missouri ranks fourth nationwide in the amount of ash produced, with sixty coal ash sites across the state handling three million tons of ash per year.[14] Most of the ash gets placed in surface impoundments or landfills, creating the potential for leaching of contaminants into groundwater or for scouring out of ponds or landfills during floods. EPA ruled that the potential for toxic elements in ash to leach from ash ponds or landfills was "sufficiently low" to prevent its classification as hazardous waste, but concerns remain that these toxic materials might end up in shallow groundwater or wells.

Pile of coal "spoils" near Germantown in Henry County *(Loring Bullard)*

Ameren operates Missouri's largest coal-fired power plant at Labadie, west of St. Louis. Each day the plant burns 130 train-cars of coal, at 100 tons per car. At peak energy production, it consumes one carload of coal in about four minutes. This prodigious rate of coal-burning goes through about ten million tons of coal per year, with 7% of that burnt coal remaining as ash. Ameren recycles much of the ash but, until recently, mixed the rest with water and pumped it out into two large

Abandoned strip pits in Henry County. Trees grow on spoil piles between the linear pits *(Jim Guldin)*

ponds on the Missouri River floodplain. Ameren was running out of room in its ponds so in 2009, the company purchased adjacent property where it intended to build a 400-acre landfill, also in the floodplain. The company applied to Missouri DNR for a construction permit, provoking an immediate outcry from neighbors and environmental groups. One local book club got re-organized as the Labadie Environmental Organization (LEO) to fight the proposal, claiming 400 members by late 2009.[15]

Ameren argued that its options were very limited. Building the landfill at a site out of the floodplain would "clog area roads with hundreds of daily truckloads of waste and cost ratepayers about $75 million per year."[16] Besides, an Ameren spokesman said, ash in the new landfill would be stored as a "dry, clay-like substance," with "just enough water to make it harden like concrete."[17] Therefore, he suggested, there will be no groundwater contamination. A retired Ameren engineer stressed that "building coal ash landfills in river floodplains has been a standard practice at hundreds of facilities across the U.S. over the last fifty years," and can be done "without any detrimental health or environmental impacts."[18]

Ameren coal-fired power plant near Labadie is Missouri's largest. *(St. Louis Post Dispatch)*

Others vehemently disagreed with that statement, including members of LEO. Tempers flared at public meetings. Even public officials were skeptical of Ameren's plan. A Franklin County Commissioner in 2011 said, "I don't think that any landfill belongs in a floodplain."[19] An "Our View" column in the St. Louis *Post Dispatch* echoed his sentiment: "Putting a coal ash landfill in the direct path of future floodwaters on one of the mightiest rivers in the world is simply bad public policy."[20] In fact, ash ponds at Ameren's Labadie and Meramec plants were "overtopped by floodwaters" in 2013 and 2015, although the company claimed the resulting discharges had "no significant adverse impact."[21]

DNR eventually granted the permit, but incorporated changes from new EPA rules published in 2015. These required the base of the landfill to be five feet above the nearest groundwater aquifer, and monitoring wells had to be installed around its perimeter. Ameren announced in 2018 that over the next few years, it would close its remaining ash storage ponds and "switch entirely to dry landfills."[22] The utility also promised to quit using coal, eventually. Recently, Ameren emphasized "robust investments in renewable energy over the next couple of decades," while also claiming it intended to retire three of its coal-fired plants by 2030.[23]

The handwriting is clearly on the wall for coal-dependent power generation. A Bloomberg report from February 2023 suggests that nearly all existing U.S. coal-fired plants "require more cash to operate than the cost of replacing them with new wind or solar projects."[24] Solar now accounts for only 1.4% of the electrical energy produced in Missouri, but its share is growing.[25] Industry representatives predict that by 2050, solar will supply 45% of our electrical needs.[26] All renewables are gaining traction in Missouri. In 2012, Missouri voters approved a mandatory renewable portfolio standard, replacing the previous voluntary goal, requiring investor-owned electric utilities to meet 15% of sales from renewable sources, with solar accounting for at least 0.3%. Renewables made up 14% of the total in-state electrical generation in 2023, exceeding 10% for three years in a row.[27]

In 2023, I toured the old coal-mined lands in Henry County. They looked a lot different than they did when I lived there. The pits were no longer in exposed, barren areas like they were in the 1960s. Back then, there were wide expanses of broken, rocky soil and steep-sided pits with gullied slopes, black with coal residue. Today, most of these areas are completely revegetated, with thick stands of small to medium-sized trees, some even growing atop spoil piles thirty or forty feet high. The water is still there, filling long, narrow lakes. It is no longer clear, but greenish with algae. Along the banks are profuse growths of cattails and sedges. In short, the old, mined lands today more nearly resemble hummocky forests and farm ponds than the stark moonscapes of fifty years ago.

Some coal strip pits have been converted to suburban real estate *(Loring Bullard)*

Water quality in the pits is probably better than it used to be. Presumably, acid from the spoil piles has mostly leached out by now or remains locked in minerals held in place by roots. Some of the vegetation I saw, like thick stands of bluestem grasses, was undoubtedly part of reclamation efforts, as were the clusters of non-native white pines growing on some spoil piles. Missouri's coal miners, through their actions, had inadvertently created "novel ecosystems."[28] The unnatural features we see today—the hills and pits—were never part of the Missouri landscape before mining, of course, but after mining became a blank slate upon which nature has written a new story, with mined over lands now hosting a unique assemblage of plants and animals—some native, some not.

South of Montrose in Henry County, I saw another use for the old mined over lands—rural housing developments. This rough ground has been turned into desirable real estate, with private access to fishing and boating lakes in what were once mining pits. In the backyards of several homes, docks have been constructed on the banks of the old pits, with pontoon boats tied up or kayaks or canoes stacked nearby. These days, this kind of property, reasonably close to the essential services of cities like Clinton, might be more valuable than it would have been for agricultural purposes.

While I was in the area, I decided to drive by to see the Montrose Power Plant, originally built and operated by the Kansas City Power and Light Company. I had consulted my Google Earth images before I left home to make sure I could remember how to get there. On the satellite images, taken in 2020, I saw the power plant and the shadows of its stacks. Just west of it was a large, dark-colored rectangle, which I took for the ash disposal area. When I neared the site in my vehicle, however, I became confused. At the place where I thought the power plant should be, I saw no buildings, just a locked gate and tiny security shack, with a big field beyond. Suddenly, I recognized the place, and knew this field was where the plant had once stood. To the west, semi-trailer trucks were passing through another monitored gate toward a huge mound of earth where the ash disposal area had been.

Only later, sitting at home, did I uncover the rest of the story. The power plant had ceased operation in 2018, and had been removed sometime after March 2020, when my Google Earth images were taken. In January 2015, Kansas City Power and Light (KCPL) announced that the Montrose power plant would be retired. The three power-producing boiler units were taken offline between 2016 and 2018. By this time, KCPL had merged with another power company, becoming the bi-state Evergy Corporation, a subsidiary of Great Plains Energy. Evergy pledged to have "net-zero" carbon emissions by 2045, which included removing from service less efficient and more polluting coal-fired plants like the one at Montrose. Power was still being produced at larger, cleaner, coal-fired plants, as well as at a 1,150-megawatt nuclear facility in Kansas built in 1985.[29]

The construction work I saw to the west of the former plant site was part of the new coal combustion residual landfill. An engineered "cap" of compacted earth was being placed over the coal ash, and below it a leachate collection system had been installed so that contaminated seepage water could be captured and treated. Around the perimeter of the large mound, groundwater monitoring wells had been installed. All of this work was part of a "post-closure plan" agreed to by Evergy in 2021.[30]

Today, threats from coal mining have decreased significantly. Most of the old mining lands in Missouri have been reclaimed and are no longer leaching acids into streams. Cleaner fuels and air ordinances mean smoke has mostly disappeared over St. Louis, Kansas City, and other cities. Atmospheric pollution from coal burning, however, is still prevalent, a shift from suffocating black smoke to invisible greenhouse gases. Missouri and mid-America are responsible for a disproportionate share of this pollution. But our state, like others, is reluctant to take actions that might increase the price of energy, or threaten jobs, or impair the economy. The old struggles have not gone away—only changed in form.

Ash landfill under construction near Montrose *(Jim Guldin)*

SOURCES

1 *Louisville Courier Journal*, April 13, 2016; courier-journal.com/story/tech/science/watchdog-earth/2016/04/13/Peabody-bankruptcy-remembering-big-hog/82980882/

2 *St. Louis Globe Democrat*, March 4, 1932, p. 4

3 *St. Louis Star and Times*, May 11, 1912, p. 4

4 *St. Louis Star and Times*, November 28, 1939, p. 5

5 *The Kansas City Star*, March 7, 1941, p. 3

6 *The Kansas City Star*, October 9, 1945, p. 1

7 Joel A. Tarr and Carl Zimring, "The Struggle for Smoke Control in St. Louis," in *Common Fields: An Environmental History of St. Louis*, St. Louis, 1997, p. 218

8 *The Kansas City Star,* August 30, 1970, p. 6

9 Dale W. Blevins, "Hydrology and Water Chemistry of an Abandoned Surface Coal Mine, Southwestern Henry County, USGS Water Resources Investigation Report 90-4047, Rolla, 1991, p. 1

10 *Springfield News-Leader,* April 23, 1990, p. 56

11 Mike Mueller and Van Beydler, "Coal Mines Water," *Missouri Resources*, Vol. X, No. X, 2014, p. 14

12 Global Energy Monitor; gem.wiki/Missouri_and_coal

13 Nebraska Public Media, "Midwest Coal-Fired Power Plants," Jan. 11, 2022; nebraskapublicmedia-06/e/news/-articles/midwest

14 EarthJustice, May 3, 2023; earthjustice.org/feature/coal-ash-states/mo

15 *St. Louis Post Dispatch*, November 11, 2009, p. 1

16 *St. Louis Post Dispatch,* January 4, 2015, p. 5

17 *St. Louis Post Dispatch,* October 1, 2011, p. 6

18 *Springfield News-Leader*, January 26, 2014, p. 1

19 *St. Louis Post Dispatch,* October 1, 2011, p. 6

20 *St. Louis Post Dispatch*, March 31, 2014, p. 9

21 *St. Louis Post Dispatch,* February 18, 2018, p. 6

22 *St. Louis Post Dispatch,* February 18, 2018, p. 6

23 *St. Louis Post Dispatch,* June 24, 2022, p. 5

24 *Springfield News-Leader,* October 13, 2016, p. 6

25 Solar Energy Industries Association, seia.org/state-solar-policy/Missouri-solar/

26 Connor O'Neil and Harrison Dreves, "Building a Solar Powered Future;" nrel.gov/news/program/2022/building-a-solar-powered-future.html

27 U.S. Energy Information Administration; eia.gov/state/analysis.php?sid=MO

28 YouTube, "How Coal Shaped Missouri's Landscapes," youtube.com/watch?app=desktop&v=Cmqjsy-pqdk

29 Evergy, "Sustainability;" evergy.com/smart/environmental-impact-link/sustainability-hub#

30 Evergy, "Post Closure Plan, Montrose Generating Station CCR Landfill;" evergy.co/-/media/files/ccr/evergy/Montrose-generating-station/ccr-landfill

CHAPTER 5: LAYING WASTE

"One of the first conditions of happiness is that the link between man and nature shall not be broken."

LEO TOLSTOY

It happened sometime in the 1980s. I don't remember the date, but I do remember the place—a small cubicle at the regional Health Department office in Springfield. I met with a tall, thin man who asked me about my attendance at the Water and Wastewater Technical School in Neosho, Missouri—when I was there; what buildings my classes were in; if I had worked in the school's laboratory. He went over my health history in some detail, then zeroed in on time I had spent in the laboratory and in one building at the school. At one end of this building, outside, was a cylindrical metal tank. It gradually dawned on me that the tank and its contents were the primary reasons for his questions, and his concern. The tank, it turns out, held an industrial waste, and that waste contained dioxin.

I left that interview with an oddly unsettled feeling. Although I felt healthy enough, a vague sense of foreboding swept over me. Was it possible I had handled some of the tank's contents in the lab? I thought about the past—about times when I hadn't felt well. Could dioxin have been the cause? What about the future? Would I get cancer? After the interview, I heard nothing more from the Health Department. Eventually, I quit worrying about it. I didn't realize that other people, at that very moment, were extremely worried about dioxin. I was slow to recognize a nationwide awakening—a reckoning with previously unknown, unseen toxins.

In 1977, Michael Brown, a young reporter for the *Niagara Gazette*, went to a meeting to learn about organized crime's involvement in the local waste disposal business. During the meeting a young woman, Karen Schroeder, began crying, pleading, "Don't let them do to you what they did to us."[1] Later, Brown interviewed her at her home, in a small community

called Love Canal, near Niagara Falls. He learned that in 1968, she had given birth to a child with severe birth defects—deafness, deformed ears, cleft palate, hole in the heart, intellectual disabilities. It was a heart-rending story, of course, but the young mother's terrible saga wasn't over.

In the wet winter of 1974, Karen looked out her back window to see that her in-ground swimming pool had partially popped out of the ground. The next summer, the family decided to replace the pool. When they pulled it out, they saw in the gaping hole strange yellowish and blue liquids with a strong chemical smell. Over the summer, the liquids ate the bottoms out of redwood posts in Karen's yard, causing the fence to collapse. Most of her bushes and garden plants died. In talking to neighbors, she learned there were other people having similar problems.[2] She told Michael Brown all about it.

In 1979, Brown published *Laying Waste: The Poisoning of America by Toxic Chemicals.* Love Canal became the ugly poster child of the movement to stop the indiscriminate dumping of toxic waste. President Carter declared Love Canal a federal health emergency and directed the Disaster Assistance Agency to investigate the site and clean it up, marking the first time federal emergency funds had been used for a man-made disaster. 950 families had to be relocated. The complex cleanup cost $400 million over 21 years.[3]

A hazardous waste dump devastated the community of Love Canal in New York State *(Bowdoin.edu)*

Sampling showed a veritable potpourri of toxic chemicals at Love Canal, including benzene, chloroform, toluene, polychlorinated biphenyls (PCBs) and dioxins. Dioxins are by-products of many industrial processes involving heat, and can also be created by incineration of solid waste, smoking cigarettes, and the bleaching of paper. They are prevalent in foods cooked at high temperatures. After ingestion, they tend to build up in fat, or bioaccumulate. They are also persistent in the environment, taking over ten years to dissipate. Their potency is legendary. Newspapers in the 1970s commonly reported that dioxin was "thousands of times more poisonous than cyanide."[4] More recent studies revised estimates of toxicity, but dioxins are still very dangerous.

Spurred to action after the Love Canal disaster, Congress passed the Comprehensive Environmental Response, Compensation, and Liability Act in 1980, allowing the federal government to directly intervene when toxic wastes are being released into the environment, or if there is a threat of release. Hazardous waste sites with the greatest potential to affect human health are placed on the National Priority List, to be first in line for cleanup. The law also established liability for companies that produce or handle hazardous wastes, and imposed a tax on chemical and petroleum industries to pay for cleanups in cases where responsible parties can't be located, or can't pay. This fund grew to about $1.6 billion in the first five years, inspiring the program's nickname—Superfund.

The Superfund story in Missouri began with dioxin. Like that of many toxic chemicals, dioxin's story goes back to a war. In the 1960s, the U.S. military sprayed herbicide over much of Vietnam to kill vegetation where the enemy could hide, and to a lesser extent to destroy his crops. The most widely used herbicide was called Agent Orange because it was stored in army-green drums bearing an orange stripe. It is a mixture of two chemicals, 2,4-dichlorophenoxyacetic acid (2-4-D) and 2,4,5-trichlorophenoxyacetic acid (2-4-5-T). One of the most toxic chemicals in Agent Orange, dioxin, is unintentional—a residual contaminant of the manufacturing process. Agent Orange was sprayed over a quarter of southern Vietnam, destroying 500,000 acres of crops. The Red Cross estimates that it poisoned three million Vietnamese families, leading to at least 150,000 children born with birth defects.[5]

In the 1960s, Agent Orange was produced by the Hoffman-Taff chemical company in Verona, Missouri.[6] The manufacturing process includes distillation, where chemical mixtures are separated by heating in a tall, cylindrical tank. Lighter chemicals rise to the top, where they are pulled off. Heavier substances are taken out from lower levels. The thickest stuff remains at the bottom of the still, the so-called "still bottoms." In the manufacturing of Agent Orange, this thick, oily bottom residue, where dioxins are concentrated, is a waste material to be hauled off, ideally to an approved treatment facility.

Dioxin contamination made Times Beach into a "ghost town." *(State Historical Society)*

After a wider recognition of its dangers, Agent Orange was banned in the United States in 1971. A few years earlier, with its primary product set to be phased out, Hoffman-Taff faced bankruptcy. In 1969, the Syntex Agribusiness Corporation bought the failing company and leased a portion of the Verona plant to the Northeastern Pharmaceutical and Chemical Company (NEPACCO) to produce hexachlorophene, an antiseptic widely used at the time in the medical industry. The distillation process used to create hexachlorophene also created dioxins. NEPACCO operated the plant from 1969 to 1971 when it, too, faced financial difficulties, mainly because hexachlorophene was also slated to be banned—after thirty-two infants in France died from high levels of the chemical in baby powder.[7]

Facing cash flow problems, NEPACCO canceled its expensive contract with the waste disposal firm that had been transporting the still bottom wastes to New Orleans for incineration. The company then contracted with Independent Petrochemical Corporation (IPC) to have the wastes removed and treated elsewhere. IPC had little experience with waste disposal, so it sub-contracted the job to a firm owned by Russell Bliss of west St. Louis County, Missouri. Between February and October 1971, Bliss hauled off six truckloads of still bottoms, nearly 18,500 gallons. He took most of the waste to his storage facility in Frontenac, where he mixed it with used motor oil. Bliss would later testify that NEPACCO didn't tell him the material was hazardous. That contradicted the testimony of William Ray, manager of NEPACCO's Verona plant, who said he had asked one of Bliss's drivers where the waste was going. According to Ray, the driver said they were putting it on a horse arena. "God, you can't do that!" Ray purportedly yelled. "You can't put that material where people are going to be walking around on it."[8]

But that is exactly what Bliss did. He used some of the dioxin-tainted waste oil for dust control on a horse arena at his own farm. Visitors were impressed with how well it worked, and soon asked Bliss to treat their stables and arenas. In May 1971, the owners of Shenandoah Stables at Moscow Mills, northwest of St. Louis, paid Bliss $150 to spray about 2,000 gallons of waste oil on their arena. The wastes gave off a very pungent, eye-watering odor. Before long, birds dropped dead from the rafters. Horses developed big sores and lost their hair. Two of the owner's daughters became ill, with abdominal pain and nosebleeds.[9]

The stable owners blamed Bliss, who insisted the material was just waste oil, and non-toxic. At their own expense, the owners removed the top 18 inches of oil-treated soil. But horses using the arena still got sick. After several months, over 60 horses had become so weak and emaciated that they had to be euthanized. Bliss was also hired to spray at Timberline Stables near Jefferson City. Twelve horses died there, and children exposed to the soil developed chloracne, a painful skin rash.[10] These owners also removed the top layer of sprayed soil, as did the owners of Bubbling Springs Ranch, another property Bliss sprayed. The contractor hired to haul off the soil from the Bubbling Springs ranch took some of it to his own home and also to the house of a neighbor. All these sites would eventually be involved in a Superfund cleanup.

Sign at the Meramec River Bridge at Times Beach, 1987 *(Smithsonian Magazine)*

After the human illnesses at stables in eastern Missouri, physicians contacted the Centers for Disease Control (CDC), which in August 1971 began sampling soil from the arenas. Having just been formed the previous year, EPA at first wasn't brought into the case. Samples of soil taken by the CDC contained high levels of trichlorophenol and polychlorinated biphenyls (PCBs), but also 30 ppm of dioxin.[11] At the time, little was known about the effects of dioxin on humans, but the deaths of so many animals were alarming. When questioned, Bliss said he didn't know where the dioxin might have come from. But after a short investigation, the CDC connected him to NEPACCO, the only company in Missouri producing trichlorophenol.[12]

When Syntex Agribusiness bought the Verona facility from Hoffman-Taff in 1969, it inherited tanks and lagoons filled with dioxin-contaminated wastewater. 255,000 gallons of wastewater were put into four unlined lagoons at the Verona plant. DNR officials later speculated that dioxin might have seeped from these lagoons into the groundwater. Wastes at the plant had also been dumped into trenches, on sloping ground only 200 feet from the Spring River. Missouri DNR in 1981 issued a health advisory against eating fish from the Spring River, which by then contained "small" amounts of dioxin.[13]

Much of the waste material was transported off-site. James Denney, a Syntex employee, told EPA that NEPACCO had paid him $25 a truckload to dump barrels of still bottom wastes on his farm, about seven miles from the Verona plant. He claimed he had no idea what was being dumped. Barrels of waste were also buried on several other farms in the area. One Syntex employee was walking on top of barrels when a loose lid gave way, submerging his leg up to the knee. The wastes "ate away his shoe and the bottom portion of his pants leg like acid." A total of 90 barrels of dioxin-containing wastes, corroding and leaking, were later unearthed at the farms. Eleven of the barrels contained wastes with dioxin concentrations as high as 2,000 ppm.[14] The CDC at the time considered any dioxin level over 1 part per billion a threat to human health, so the barrel contents were two million times the CDC's health advisory level.

Waste sludge from the lagoons at Verona and contaminated soil from the farms were shipped to the Denney Farm for treatment in a mobile incinerator. This $5 million truck-mounted incinerator, nicknamed the "Blue Goose" because of the long, neck-like conveyor belt at one end, had been tested by the EPA in the early 1980s. Originally designed to burn oily sand from beachfront spills, the Blue Goose had been modified to burn dioxin, and after testing was deemed successful.[15] The Goose landed on the Denney Farm in December 1984 and would stay there for the next three years, burning an estimated 10 million pounds of dioxin-laden wastes from seven "dioxin danger spots" in southwest Missouri.[16]

In the 1970s, CDC and EPA officials thought that dioxin in soil would dissipate within a few years. But when they retested the soil at Shenandoah Stables in 1982, they found dioxin levels to be about the same as when they tested in 1971. The EPA advised closing the stables and prepared to start cleanup operations. But before the cleanup could begin, the Environmental Defense Fund published a leaked EPA document listing 14 confirmed dioxin-contaminated sites in Missouri. The town of Times Beach was on the list. Seeing this for the first time in the newspapers, the residents of Time Beach were angry and dismayed, feeling that the EPA had betrayed them by not notifying them sooner.[17]

Flooding in Times Beach in 1982 eroded dioxin-contaminated soil. *(Environmental Protection Agency)*

Times Beach, on the Meramec River west of St. Louis, was platted in 1925 as a summer resort. The land, originally owned by the St. Louis *Star Times*, was subdivided into 6,000 lots, 20 by 100 feet each. Single lots sold for $67.50 with a subscription to the newspaper. The town grew slowly at first, but by 1935 there were 1,300 property owners, "90 percent of whom were St. Louisans."[18] People came for boating, fishing and swimming in the river, but also for the dance halls and taverns. Most of the town was in a floodplain, and many residents built their cabins and houses on stilts. Overbank flooding from the nearby Meramec River was a chronic problem, causing many lot owners to eventually abandon the enterprise. In 1957, 3,285 lots were auctioned as part of a delinquent tax sale.[19]

In June 1972, Times Beach paid Russell Bliss $2,400 to spray the town's twenty-three miles of dirt streets for dust control. Over six years, Bliss sprayed almost 250,000 gallons of used oil.[20] After the well-publicized problems at Verona and the horse stables, the EPA knew

Huge incinerator built at Times Beach burned 265,000 tons of contaminated soil and materials. *(Fox 2 News)*

to test the sprayed soil at Times Beach. They wrapped up extensive testing there near the end of 1982, finding dioxin concentrations as high as 1.3 parts per million along the town's entire road network.[21] Just as sampling was wrapped up, a historic flood occurred on the Meramec, reaching a crest of fourteen feet above flood stage and inundating the town. Residents had to be evacuated.

On December 23, 1982, with sampling results in hand, the EPA recommended that the town not be reinhabited. Officials warned that more flooding would further spread the contamination, and they were unsure of the potential health effects. Signs were hung on the ends of the Route 66 bridge at the edge of town warning motorists to "stay in your car, keep your windows closed." The deserted town of Times Beach was added to the Superfund National Priorities List (NPL) in 1983, putting it in line for a herculean cleanup effort.[22]

EPA contractors installed an incinerator at Times Beach in 1996. The agency had learned from the Blue Goose that incineration effectively destroyed dioxin. But the Goose was too small and too slow for the massive amount of soil needing treatment at Times Beach, so the incinerator assembled there was huge in comparison. It had a 75-foot-long, 14-foot-diameter "countercurrent, controlled-air, rotary kiln" and an "oxygen-enhanced" secondary burn chamber, which at 1,750 degrees Fahrenheit was considered hot enough to destroy any remaining contaminants. A covered conveyor belt fed not only soil, but also shredded insulation, flooring, carpet, mattresses, and brush into the incinerator. When the burn was completed in June 1997, over 265,000 tons of soil and materials had been incinerated, at a total cost of $118 million.[23]

Obviously, the ordeal at Times Beach was traumatic for the residents, who were not only potentially exposed to deadly toxins but also lost their homes. Discussions of a possible buyout of their properties occurred in early 1983 when President Reagan appointed a Times Beach Dioxin Task Force. In February 1983, the EPA announced that the agency would pay $33 million of the expected $37 million cost to buy 800 residential properties and 30 businesses.[24] The Federal Emergency Management Agency (FEMA) coordinated the relocation of residents. But even after the buyout, the travails of the former residents weren't over, as many of them struggled to find financial assistance to rebuild their lives.

During the buyout negotiations, EPA and Missouri officials squabbled over who would take title to the vacated town, with its potential liabilities. The state of Missouri finally acquired the land in 1999. Soon afterwards, Route 66 Park was opened on the abandoned and cleaned up site of Times Beach, with a visitor center in a former commercial property at the east end of the Meramec River bridge. By 2001, the townsite had been removed from the NPL, but its story lived on in popular media. *Life* magazine ran an article about it, the "Ordeal of a Poisoned Town." Others put a lighter spin on it. Johnny Carson joked on *The Tonight Show* that he would rent a "glass-bottomed Winnebago" to tour Times Beach.[25]

Not surprisingly, there were literally thousands of lawsuits generated by the dioxin contamination in Missouri, against NEPACCO, the hexachlorophene manufacturer; Independent Petrochemical Corporation (IPC), which sub-contracted with Russell Bliss; and Bliss, the waste hauler. The owners of Shenandoah stables collected $10,000 from Bliss in 1976 and $65,000 from NEPACCO in 1981. IPC paid out $1 million each to the two daughters of the arena's owners. In 1992, 381 former residents of Times Beach settled out of court in seventeen separate lawsuits.[26]

Ripples from the dioxin contamination in Missouri spread nationwide, providing a major impetus for Congress to pass the Toxic Substances Control Act in 1976, which required extensive testing of chemicals posing an "unreasonable risk" to humans or the environment. Also in 1976, the Resource Conservation and Recovery Act was passed, regulating the transportation and disposal of hazardous waste. A key feature of this law is its "cradle to grave" requirement, meaning that paperwork must follow hazardous wastes from where they were first created to where they finally end up.

Superfund was used to clean up many more hazardous waste sites, including several in Missouri, but for most of its existence the program faced severe criticism. An editorial in *The Kansas City Star* charged that "one of Superfund's biggest faults is how it has been much more efficient at empowering government bureaucrats and enriching lawyers and consultants than cleaning up the environment."[27] In 1995, Congress terminated the tax on chemical and petroleum companies for the cleanup trust fund. With no new corporate taxes coming in, the trust fund rapidly became depleted. The EPA had to ask Congress for additional funding for cleanups when responsible parties couldn't be found, or forced to pay. For many environmentalists, this changed the program's original "polluter pays" premise to one of "taxpayer pays," deleting, to them, a "core principle" of the law.[28]

The tax on chemical manufacturers for the trust fund was reinstated in November 2021 as part of the Infrastructure Investment and Jobs Act, once again providing funds for times when polluters couldn't pay. Over the years, however, the EPA has done a pretty good job of locating and leaning on Potentially Responsible Parties (PRPs). Since Superfund's inception, about 70% of costs have been paid by PRPs. In total, about 40,000 sites across the U.S. have been involved in the Superfund program, with over 1,300 making the National Priorities List. As of August 2022, about 450 of these sites had been delisted, meaning that in most cases, cleanups have been successfully completed.[29]

Dioxin put Missouri on the Superfund map, but since 1980, when the program was initiated, our state has added many more sites with a variety of problems. Of the 30 sites currently on the NPL, most are old industrial sites. At the Bee Cee site in Malden, hexavalent chromium was left over from the manufacturing of aluminum door and window frames.[30] At Quality Plating in Sikeston, residual chromium, nickel, lead, and zinc from electroplating contaminated groundwater.[31] Lee Chemical, on the south edge of Liberty, stored solvents awaiting disposal in an old water treatment plant. The site became contaminated with dichloroethene and trichloroethene, chemicals similar in nature to the toxic solvent trichloroethylene (TCE).[32]

At the Lake City Ammunition Plant near Blue Springs, volatile organic chemicals (VOCs), polycyclic aromatic hydrocarbons (PAHs), heavy metals, and depleted uranium were left from the manufacturing and testing of small caliber ammunition.[33] At Armour Road in North Kansas City, a former herbicide mixing facility contaminated soil and the Missouri River alluvial aquifer with arsenic.[34] The old Fulbright Landfill, north of Springfield, leached cyanide, heavy metals, and VOCs from municipal waste into groundwater.[35] Also at Springfield, a waste lagoon serving Solid States circuit board manufacturing company leaked TCE into the karst groundwater system.[36]

The Pools Prairie complex in southwest Missouri is an example of a site containing many different pollutants and several separate areas in need of remediation. In the early 1940s, the U.S. government acquired 43,000 acres south of Neosho for construction of the Fort Crowder Army Base. After the war, in the 1950s, a portion of the base was turned over to the Air Force to manufacture rocket engines, and later to a private company producing jet airplane engines. Contamination in four separate areas included TCE and a variety of VOCs.[37]

Lake City Ammunition Plant near Blue Springs *(U.S. Army Joint Munitions Command)*

The Bannister complex in Kansas City is another sprawling, difficult site. Beginning in the 1950s, this was home of the Bendix Corporation, which had a U.S. government contract to build non-nuclear parts for nuclear weapons. Other companies eventually took over the complex, leaving almost 800 different chemicals in soil and groundwater, including PCBs, TCE, depleted uranium and asbestos.[38] At a town hall meeting in 2010, residents told stories about friends and coworkers who they believed had "suffered and died" from working there.[39] The site was nominated for the Superfund NPL, but in 2011 EPA Administrator Karl Brooks said that because of the red tape associated with priority listing, keeping it off the list would allow the cleanup to begin sooner.[40]

Almost fifty years ago, I was possibly exposed to dioxin at the Water and Wastewater Technical School in Neosho. It was only recently, when researching for this book, that I found out how the tank of dioxin-laden waste happened to end up there. In 1971, NEPACCO was looking to get rid of their still bottom wastes from the plant at Verona. The company first approached the nearby city of Aurora to see if it would accept the wastes at its wastewater treatment plant. The city refused. NEPACCO then contacted the city of Neosho. At the time, Neosho's wastewater plant was being operated by the Water and Wastewater Technical School. The school administrator later claimed that NEPACCO did not inform him the wastes contained dioxin, but the EPA maintained he should've known the material was hazardous, since it came from the manufacturing of a toxic chemical.[41]

Regardless, the technical school signed an agreement with NEPACCO in October 1971. The school was paid 2 ½ cents per gallon to accept, treat, and dispose of the dioxin-laden waste. Apparently, the plan was to send it through Neosho's sewage plant for treatment. Most of it was stored there in an old, unused concrete digester tank. There are no records to show that any of it ever received treatment. Neosho city officials claimed they were not told what was in the tank when they resumed operating the treatment plant in 1974.[42] Three years later, city workers filled the old digester tank with gravel, broken asphalt, and dirt. The dumping action sloshed much of the waste out of the tank. Assistant Plant Superintendent Gary Rice said it created a "wave of oozing sludge" with the "consistency of chocolate pudding." Rice admitted that he and his employees had all "been in direct contact with it."[43]

Some of the waste had also been placed in a storage tank at the technical school, presumably to be "studied" by industrial waste treatment classes. But EPA records show the only contact students had with the tank was to "practice opening and closing the valve," spilling wastes onto the ground.[44] In early 1981, the EPA removed the tank from the school and placed fifteen 55-gallon drums of contaminated soil into an "old army bunker" on the former Fort Crowder Army Base. The technical school, which had received federal support, closed in December 1981, leaving the EPA "among its creditors."[45]

The history of Superfund is in many ways a tragic saga. Each of the 1,300 sites on the National Priority List represents a heartbreaking chapter in the story, holding large measures of fear, anger, and mistrust. Another part of the tragedy is that Superfund was so mismanaged; politics were played and sweetheart deals made, thus diverting time and money away from real, potentially deadly problems. When money went to lining the pockets of consulting or cleanup firms, sites that should've been cleaned up weren't. For many sites on the list, serious risks—proverbial ticking time bombs—remain.

On the positive side, many heavily contaminated sites have been cleaned up and thus, future problems have been averted. Hazardous materials laws of the 1970s and 1980s, like the Toxic Substances Control Act, the Resource Conservation and Recovery Act, and CERCLA, or Superfund, should help prevent the creation of more contaminated sites like Times Beach or Verona. But there are hundreds of sites that still need to be cleaned up. It's a monumental, expensive job, but it's also an investment in the future health and safety of American citizens, because each of those sites poses ongoing, unnecessary risks for those potentially exposed to the world's most dangerous substances.

SOURCES

1 *The Los Angeles Times,* May 25, 1980, p. 75

2 Michael Brown, *Laying Waste: The Poisoning of America by Toxic Chemicals,* New York, 1979, p. 5

3 USEPA, website: "Love Canal, Niagara Falls, Cleanup Activities;" cumulus.epa.gov/supercpad/SiteProfiles

4 UPI Archives, March 20, 1982, website: upi.com/Archives/1983/03/20Dioxin-a-ticking-time-bomb

5 United States Institute of Peace, website: suip.org/publications/2022/10/addressing-harmful-legacy-agent-orange-vietnam#

6 Toxicsites.us/site.php?epa_idMOD007452154

7 *The New York Times*, August 29, 1972, p. 10

8 *St. Louis Post Dispatch,* January 25, 1983, p. 1

9 *Springfield Leader and Press*, October 27, 1982, p. 3

10 *St. Louis Post Dispatch,* August 27, 1974, p. 1

11 *Springfield Leader and Press*, October 27, 1982, p. 3

12 *Springfield Leader and Press*, May 23, 1975, p. 18

13 *St. Louis Post Dispatch*, December 12, 1982, p. 79

14 *Springfield Leader and Press*, February 3, 1980, p. 8

15 *St. Joseph News-Press*, August 5, 1987, p. 9

16 *Springfield Leader and Press*, March 13, 1983, p. 10

17 William Powell, "Remember Times Beach: The Dioxin Disaster, Thirty Years Later," in *St. Louis* Magazine, December 3, 2012, accessed at: stl.mag/com/Remember-Times-Beach-The-Dioxin-Disaster-30-Years-Later/

18 *St. Louis Globe-Democrat*, April 16, 1935, p. 14

19 *St. Louis Post Dispatch*, June 25, 1957, p. 10

20 *St. Louis Post Dispatch*, February 16, 1983, p. 12

21 *Springfield News-Leader*, August 3, 1984, p. 4

22 *St. Louis Post Dispatch*, February 23, 1983, p. 1

23 *St. Joseph News-Press*, April 13, 1997, p. 15

24 *The Kansas City Times*, March 17, 1983, p. 44

25 *St. Louis Post Dispatch*, March 10, 1985, p. 41

26 *St. Louis Post Dispatch*, November 5, 1992, p. 3

27 *The Kansas City Star*, January 26, 1994, p. 47

28 *The Kansas City Star*, May 14, 1998, p. 74

29 Website: en.wikipedia.org/wiki/List_of_Superfund_sites

30 USEPA website: cumulus.epa.gov/supercpad/cursites/ctsitinfo.cfm?id=0701439#

31 USEPA website: cumulus.epa.gov/supercpad/SiteProfiles/index.cfm?fuseaction=second.cleanup&id=0701442#

32 USEPA website: cumulus.epa.gov/supercpad/cursites/csitinfo.crm?id=0701378&msspp=med

33 USEPA website: cumulus.epa.gov/supercpad/cursites/csitinfo.cfm?id=0701757

34 USEPA website: cumulus.epa.gov/supercpad/cursites/csitinfo.cfm?id=0702515

35 USEPA website: cumulus.epa.gov/supercpad/cursites/csitinfo.cfm?id=0701153

36 USEPA website: cumulus.epa.gov/supercpad/SiteProfiles/index.cfm?fuseatcions=second.cleanup&id=070139

37 USEPA website: cumulus.epa.gov/supercpad/cursites/csitinfo.cfm?id=0702918

38 USEPA website: epa.gov/mo/bannister-federal-complex-kansas-city-missouri

39 *The Kansas City Star*, April 22, 2010, p. 11

40 *The Kansas City Star*, September 2, 2011, p. 4

41 *St. Louis Post Dispatch*, December 12, 1982, p. 79

42 Ibid.

43 Ibid.

44 Ibid.

45 *The Kansas City Star*, October 15, 1982, p. 5

CHAPTER 6: HEAVY METAL WITH A WEIGHTY PAST

"The care of the earth is our most ancient and most worthy and, after all, our most pleasing responsibility."

WENDELL BERRY

Not far from our home in Springfield, on a rise above Pearson Creek, is a large pile of dirt and broken rock. This steep-sided mound has been there for quite some time, as evidenced by the size of the trees growing on it. Nearby are massive concrete pylons, studded with heavy bolts; apparently anchor points for a large piece of equipment that once stood there. These remains mark the site of the Badger Mill, part of the Badger Lead and Zinc Mine, which operated here in the early 20th century. Hidden behind a residential swimming pool, this overgrown spot represents a nearly forgotten piece of history. But it's not hard to imagine the noise and dust that once attended the workings here, as men labored to dig, raise, haul, dump and grind heavy loads of lead ore. In 1907, 30-year-old Elmer Dickensheet was in a "tub" being lowered into a shaft 250 feet deep when the cable broke, killing him instantly in the fall.[1]

It's just one tragic story among many in an inherently dangerous occupation. Lead, with its wide variety of commercial and military uses, has been mined for at least 3,000 years, much of it by slaves working under horrible conditions. Soft and malleable, lead has been fashioned into water pipes, for example in Roman bathhouses, and gladiators covered their fists with lead knuckles. The Chinese and Greeks used it to make coins and the Babylonians and Assyrians manufactured lead plates for building construction. Artists loved leaded white paint because of its thickness, density, and long-lasting finish. Lead remains important to society today, widely used in batteries and shielding for X-ray equipment.

But the dull gray metal has a much darker side. The fate of the Romans is legendary, poisoning themselves with "sugar of lead" sweetener, wine from leaded goblets, and lead-based cosmetics. We now know that lead absorbed into the body can cause permanent damage to the brain and nervous system, leading to seizures, loss of muscle control, behavioral changes and learning disabilities. Lead poisoning has afflicted thousands of children who eat sweet-tasting chips of lead-based paint. It has poisoned people who drink water from lead pipes, and those who imbibe acidic food or drinks from lead-glazed vessels. Lead is two-faced—remarkably useful and versatile on the one hand, an insidious poison on the other.

Just a few miles from the remains of the Badger Mine stands a historical marker, placed at the point where explorer Henry Schoolcraft camped near the James River in January 1819. There, he saw shiny lumps of lead ore, or galena, on the stream bottom and embedded in the banks. Schoolcraft was not overly impressed with the extent and value of this deposit, the endpoint of his three-month sojourn in Missouri Territory. He was more interested in the very productive lead mines around Potosi, in the eastern part of the territory, where he arrived after a long journey from his home in New York. Surrounded by about forty mines, Potosi must have resembled a giant prairie dog village when Schoolcraft arrived, with holes and piles of soil and rock wastes, or tailings, everywhere. The air was no doubt hazy from all the smoking furnaces.[2]

Lead mining in Morgan County, ca. 1903 *(State Historical Society)*

The lead ore, principally galena, was mined as detached lumps from the reddish soil. At first, miners used only picks and shovels, but later, drilling and blasting became commonplace. Chunks of ore were fired in furnaces built of limestone blocks stacked on the side of a hill so that the bottom sloped toward a trough on the downhill side. Logs were places in the furnace, ore was piled on top, then more logs were added and a fire started. The fire was kept low at first to roast the ore and drive off the sulfur but was then allowed to burn hotly for about twelve hours, until the molten lead had all flowed into the basin in front.[3]

Despite their inefficiencies, the mines and smelters at Potosi yielded a lot of lead. Schoolcraft estimated that nine million pounds were produced there between 1798 and 1816.[4] Production had tapered off by the late 1820s, and some mining authorities forecast that lead production in the state had already peaked. But they were wrong. Significant new deposits of lead and zinc were found in southwest Missouri in the 1840s, in what would become known as the Tri-State Mining District. Twenty years later, the St. Joseph Lead Company began mining deep, rich veins of ore in what is now called the Old Lead Belt, about twenty miles northwest of Potosi. By the mid-1890s, Missouri mines produced over a million tons of lead annually, ranking the state second in production only to Colorado.[5]

Inside the Federal Mine in the Old Lead Belt *(State Historical Society)*

Smokestack at the closed smelter at Glover *(Loring Bullard)*

But lead production in Missouri had still not peaked. In the 1950s, exploratory drilling twenty-five miles southwest of Potosi located a lead deposit in a deep, ancient reef structure. This newly discovered ore body, lying about 1,200 feet below the surface, was called the Viburnum Trend because its northern end lay near the small town of Viburnum. Geologists still debate how the lead got into the reef limestone, but a prominent theory is that tectonic plates smashing together created intense heat, superheating groundwater, which dissolved metals out of volcanic rocks and injected them into pores in the reef structure. Mining in this New Lead Belt began in 1960. Production here, along with the Old Lead Belt and Tri-State District, secured Missouri's position as the top lead-producing state, by 1990 providing over 90% of the nation's lead.[6]

Today, the Viburnum Trend is the only active lead mining area in the state. Ore is removed from deep shafts and hauled to the surface, where it is milled to a fine powder. This finely ground ore is then mixed with water and sent through a chemical flotation process, where the metals—lead, zinc, and smaller amounts of copper—are separated from the rock slurry. The waste rock slurry is then pumped to ponds for settling and treatment. The metal concentrate is dried to the consistency of a "damp powder," ready to be sent to a smelter for refining.[7] At the smelter, the concentrate goes into a blast furnace. Molten lead and the unwanted slag settle to the bottom of the furnace, while lighter elements and impurities float to the surface. The molten lead is collected and refined using chemical processes to remove the remaining impurities.

At each stage of this process, from mining through smelting and refining, there are opportunities for environmental contamination. Water pumped from abandoned or working mines can have high concentrations of lead, zinc, and other metals. Tailings ponds may leak into groundwater or overflow during floods, sending lead-contaminated sediment downstream.

Milling and transporting ore or concentrate can leave lead dust and residue around mills and along roadsides. And smelting can result in air pollution that can be inhaled by nearby residents or contaminate downwind properties as fallout.

Lead bars stacked at the St. Joseph Lead Company smelter at Herculaneum, ca. 1900 *(State Historical Society)*

By the 1960s, health officials were urging that all these routes of potential human exposure be investigated. Lead mining and processing industries stressed that lead exposures in their operations were regulated by mine safety officials, and lead posed no serious health threats to the public. But they were wrong. In 1970, a local landowner, Earl Governo, noticed that the legs of his horses would suddenly buckle. The animals fell to the ground "twitching and gasping convulsively." A veterinarian with the University of Missouri clinic diagnosed the horses with lead poisoning.[8] Governo's farm was just a few miles downwind from the American Smelting and Refining Company (ASARCO) smelter at Glover, Missouri.

Subsequent investigations of the smelter site by environmental agencies revealed several areas of contamination. In 1994, ASARCO entered a consent decree with EPA to clean up contaminated soil, sediment, and surface and groundwater, which contained elevated levels of lead, cadmium, zinc, thallium, and arsenic. ASARCO undertook corrective action, including capping of slag piles, until 2004, when the company declared bankruptcy. By then, the smelter had already closed. The Doe Run Company, which purchased the ASARCO holdings, completed the cleanup after 2011 using funds from ASARCO's bankruptcy proceeds.[9]

The lead smelter at Herculaneum, built in 1892 by the St. Joseph Lead Company, also created environmental problems. Founded in 1864, St. Joe became the dominant mining company in the Old Lead Belt, with mines near Bonne Terre, Desloge and Leadwood. Hundreds of miles of underground railroads hauled ore to vertical shafts, where it was lifted 200 to 400 feet to the surface and transported by rail to the smelter at Herculaneum. Residents living near the smelter were often "half-blinded by the black snow" from it.[10]

In 1975, more serious problems came to light when the Missouri Department of Health tested blood lead levels of children living in Herculaneum. Public health experts warned that any level of lead poses a danger to children. Thirteen of the 208 children tested had blood lead levels above the health standard at the time (30 micrograms of lead per deciliter of blood). The number was lower than expected, but Dr. Mostapa Fahim, a University of Missouri medical researcher, suggested that more sensitive tests should have been used. In 1984, the Missouri Department of Health again tested the blood of children in Herculaneum. This time, almost half of the children tested had over 30 ug/dl of blood lead. Dr. Pat Phillips, Environmental Epidemiologist for the Missouri Department of Health, said the 1975 tests may have been in error due to the method used to draw blood.[11]

In the wake of these findings, the EPA tested air and soil lead levels in Herculaneum. Lead levels in air near the smelter were markedly elevated, making Herculaneum one of only two spots in the U.S. failing to meet the Clean Air Act standards for lead, the other being East Helena, Montana. Lead levels in the town's soil were also high. Over the years, most of the lead concentrate had arrived in trucks. Leaded dust fell from leaky beds, tires, and bumpers, gradually building up in the soil along roadways. Some road dust near the smelter was found to be "nearly 30% pure lead," prompting DNR to post warning signs on city streets.[12]

By this time, St. Joe Minerals (formerly St. Joe Lead Company), which constructed the smelter, had merged with Homestake Lead to become the Doe Run Company. Doe Run entered a consent decree with the EPA and Missouri Department of Natural Resources (DNR) in 1990, agreeing to reduce lead emissions in air and clean up contaminated soil in Herculaneum.[13] Ten years later, the company was forced to undertake an even larger cleanup, buying over 160 homes near the smelter and replacing the soil at 700 properties. The final chapter of the troublesome smelter began in 2010, when Doe Run agreed to close it down and pay $65 million for past violations of the Clean Air and Clean Water Acts. The smelter closed in 2013, and final demolition began in 2017.[14]

Tailings piles or ponds from lead mining can also create hazards. In May 1977, a heavy rainstorm washed out a big section of a steeply sloping tailings pile west of Desloge, allowing at least 120,000 cubic yards of mining wastes to flow into the nearby Big River. The slimy mess had the "property of quicksand." Several head of livestock got stuck in it and perished. With every flood, contaminated sediment moved farther downstream, spreading the problem over thirty miles of river.[15] Because lead accumulates in fatty fish, the Missouri Department of Conservation advised anglers not to eat carp or redhorse suckers downstream of Desloge. In 2017, proceeds from a $24 million damage assessment were used to anchor soil and revegetate riverbanks to keep the lead from moving farther downstream.[16]

After a decade of mining in the New Lead Belt, Doe Run realized that new sources of lead would soon be needed, and began to look elsewhere. Their immediate focus was on a potential extension of the Viburnum Trend, which ran north to south along the western sides of Iron and Reynolds Counties. If the trend continued farther southward, productive mines might be developed under sections of Shannon, Carter, and Oregon Counties. Much of this land was in the Mark Twain National Forest.

Big Spring flows from the Ozark Aquifer, once proposed as a "sole source" of drinking water in southern Missouri *(Gayle Harper)*

With ore bodies lying 1,000 feet or more underground, exploratory drilling was crucial for mine development. It could take up to ten years to locate new deposits, determine their value, and build infrastructure. Typically, mining doesn't begin until the price of lead makes it economically viable. Doe Run began exploratory drilling south of the Viburnum Trend in the early 1980s, but the price of lead dropped precipitously in 1984, when labor strikes and stagnant markets brought lead production to a halt. The company paused its exploratory drilling in 1986 when the price of lead "bottomed out" at 18 cents per pound.[17]

Mining in the Mark Twain National Forest generated controversy, especially in one particularly wild section of it. Environmental groups had long sought official wilderness designation for the "Irish Wilderness," a piece of remote Forest Service land abutting the Eleven Point River in Oregon County. In 1983, they got a bill before Congress to that end, sponsored by Missouri Senators Tom Eagleton and John Danforth. But St. Joe Lead had applied to the Forest Service in 1979 to allow exploratory drilling in the Irish. By 1983, the company had drilled several exploratory holes near the proposed wilderness area, and a St. Joe spokesman claimed the company found "strong indications that lead would be found in quantities suitable for mining."[18]

Three U.S. Congressmen toured the area of the proposed wilderness in May 1983. Harold Volkmer, a Democrat from Hannibal, Missouri, and John Sieberling, a Democrat from Ohio, were in favor of a wilderness designation. Missouri Congressman Bill Emerson, a Republican from De Soto, opposed wilderness status and favored mining, observing, "we're talking about deep mining, not bulldozers coming through and knocking all the trees down."[19] The dispute spread nationwide, becoming the most contested wilderness issue before that session of Congress. But less than a year later, in May 1984, Ronald Reagan signed the Wilderness Bill, creating the 16,500-acre Irish Wilderness, where mining and timber cutting were forbidden.

Lead mining in other parts of the National Forest also generated controversy. In 1988, the Forest Service reported that it had received over 1,000 comments on its draft Environmental Impact Statement (EIS) on Doe Run's mining proposal near Winona in far southern Missouri. Ron Kucera, Deputy Director of the Missouri Department of Natural Resources (DNR), charged that the EIS was "highly generalized and of extremely limited value," and suggested the Forest Service provide a "full blown EIS for each specific mining proposal." Doe Run's president, Jeffrey Zelms, reacted angrily, noting that with increased lead prices, doubling since 1986, the company was eager to get leases signed before the price of lead dropped again. On December 6, 1988, the Forest Service granted two leases for Doe Run to explore for lead by drilling on 3,700 acres of national forest land near Winona.

The Sierra Club and Missouri Coalition for the Environment immediately appealed the decision, citing the lack of an adequate EIS as required by the National Environmental Policy Act. This appeal delayed the issuance of leases. In addition, the Forest Service ruled that a separate environmental review would now be required for each specific mining proposal. An exasperated Zelms said his company simply wanted to drill, even with "no guarantee that it could proceed with mining." But Dan Leary of the newly formed Committee for a Lead-Free Ozarks suggested that Doe Run "would be able to exert enough political pressure to gain the mining rights once it gets the drilling rights."[20]

Within a few months, the Forest Service halted permits for exploratory drilling. That decision came from Larry Hensen, Associate Deputy Chief of the Forest Service in Washington, D.C., reversing the earlier ruling by regional Forest Service officials in Milwaukee. The new ruling required Doe Run to submit a mining plan before exploratory drilling leases could be approved—a necessary step, the Forest Service suggested, because a lease to drill would "automatically give Doe Run the right to extract any minerals it finds." Zelms fumed that he was "not quite sure how you could develop a mine plan without being able to drill first."[21]

Most of the mining controversies in the 1980s and 1990s were in the eastern Ozarks, but older mining areas had their own set of problems—for example, the Tri-State Mining District, at the corners where Kansas, Oklahoma, and Missouri meet. Lead ore, mostly galena, found at the surface down to a depth of about 75 feet, was first mined there in the 1850s. By 1870, Joplin, at the center of the district, had become a rowdy boom town. Thousands of mines were developed in the district, although most tapped small deposits and were exhausted within five to ten years. By the 1950s, high grade ore deposits had mostly played out.

Left over from the Tri-State mining days were hundreds of shafts and shallow pits and piles of waste rock, called chat, scattered across the landscape. Abandoned, water-filled shafts and pits could leach lead, zinc and cadmium from mining wastes into shallow groundwater and in some cases, private wells. High zinc levels in sediment damaged aquatic life in Center Creek, which receives 70% of the runoff from mining areas in the district. Air and soil were contaminated from the seventeen smelters that once operated there.[22] Much of the contamination was in the Oronogo-Duenweg Mining Site, in western Jasper County. Metals production peaked there in 1916, by which time about 123 million tons of ore had been processed to yield 40,000 tons of lead, leaving about eight million tons of mining and milling wastes.[23]

In 2003, mining areas in the Tri-State District were placed in the Superfund program. Under the guidance of the EPA, contaminated soil was removed from residential yards. If levels exceeded 800 micrograms of lead per kilogram of soil, the entire yard was excavated to a depth of twelve inches and clean topsoil brought in. Contaminated soil and tailings were taken to a local repository and stockpiled for potential reuse. Much of the waste was used as filling for roadbeds, but some was used in the production of asphalt, railroad ballast (what tracks are laid on), and roofing abrasives. The extensive cleanup effort produced the desired results. In Jasper County, the percentage of children with elevated blood lead dropped from 14% to 2% after contaminated soil was removed from the yards of 2,500 homes.[24]

Remediation of mining wastes reduced local sources of contamination, but wider gains were made when lead was removed from commonly used products, especially paint and gasoline. Lead-based house paint was banned in the U.S. in 1978. Starting in the early 1970s, new vehicles were designed to run on unleaded gasoline. Leaded gas for passenger cars was banned in the U.S. in 1996. These efforts made a remarkable difference. Blood lead levels in the U.S. dropped from an average of 15 ug/dl in 1976 to 2.7 in 1994. Still, the CDC estimates that 500,000 children in the U.S. are over the current blood lead limit of 3.5 ug/dl.[25] Significant quantities of lead-based paint remain in older, inner-city neighborhoods. Testing of children less than 5 years old in St. Louis indicates that 16 in 1,000 African-American children have blood lead levels over 10 ug/dl, while 9 in 1,000 white children do, a disturbing example of environmental injustice.[26]

Gravel tailings from lead mining near Desloge. *(Loring Bullard)*

I wanted to see for myself Missouri's lead mining areas—the Old Lead Belt, the New Lead Belt, and the Tri-State District. First, I explored them on Google Earth, then later by vehicle and on foot. Near Desloge in the Old Lead Belt, at the Missouri Mines State Historic Site, vacated structures and mining equipment remain from Federal Mill No. 3, constructed in 1907. In the old powerhouse is a small but fascinating museum, featuring underground railroad engines, dioramas of early mining techniques, and cases filled with shiny galena minerals and brilliantly-colored crystals. A map shows major mines and mills in nearby Bonne Terre, Leadwood and Desloge. A gray overprint shows the extent of underground shafts and drifts, creating a complex cobweb with hundreds of connecting passageways.

Near the old Federal Mill are many acres of tailings, with bike and hiking trails winding through the gravelly mounds. A few miles away, along the Big River, the gray "slime" that flushed into the Big River from a tailings pile is clearly visible on Google Earth images. In Bonne Terre, one can still tour an abandoned lead mine, deep underground. This gigantic, water-filled cavern is a favorite destination for scuba divers, who can explore the submerged remains of the mining era.

The New Lead Belt, or Viburnum Trend, can be delineated on Google Earth images by the linear cluster of mines and tailings ponds running from Viburnum southward. Tailings ponds for these mines are large and whitish due to light reflecting from the surface of the pale, dried tailings. From an image equivalent to a view sixty miles up, the white dots of tailings ponds are clearly visible, forming a nearly straight line—from the large Sweetwater Mine at the south end of the trend, moving northward through Fletcher, West Fork, Brushy Creek, Buick, and Casteel mines. The footprints of the mines themselves are not that large, mostly less than forty acres, but several of the tailings ponds are huge, over 400 acres. Where runoff from forested side drainages enter the mostly white ponds, the trapped pockets of water reflect an unnatural shade of turquoise.

Tailings ponds in the Viburnum Trend are clearly visible from space *(Google Earth image)*

Most of the lead has been extracted from tailings in the ponds, but residual contaminants remain, providing a potential source of contamination for both surface and groundwater. There is always the risk of dams breaching after storm events, as has happened in the past, washing contaminated sediment downstream. Wastes stored in ponds will be there for a long time, and experience has shown that it is difficult to "reclaim" these sites. In total, the tailings ponds for all the mines in the Viburnum Trend cover about three or four square miles of real estate—property that will most likely be unavailable for other uses for decades. Lead may be essential to our society, but the collateral damage from its extraction in Missouri is colossal.

With the awakening to its dangers, lead is less of a threat than it used to be. Mining and mine wastes are now better controlled. Lead is gone from gasoline, and far fewer houses contain lead-based paint. Much of the lead used in the U.S. is now recycled, about 950,000 metric tons per year.[27] Over 99% of lead-acid batteries are recycled, making them the most recycled consumer product in the world.[28]

Many of them go to the Doe Run Recycling Center, one of the world's largest, where about 8 million batteries per year are recycled.[29] All this recycling means many tons of lead that won't have to be mined.

The old lead mining district along Pearson Creek in Greene County is today a historic relic, with shaft openings and spoil piles scattered through the woods. A cleanup has been completed at the site of the old Badger Mine. Badger landed on the EPA Superfund list when DNR conducted a survey of old mining sites in 2007. Soil near the old mine had lead concentrations of over 3,000 parts per million, with 400 parts per million considered the limit for potential human exposure.[30] With EPA funding, clean topsoil was brought in to cover contaminated soil. The old spoils pile which, to the relief of residents, contained very little lead contamination, still looms over the swimming pool. But some problems linger. A few years ago, portions of a house collapsed into an old mining shaft less than a mile from the Badger mine site.

As one of the largest producers of lead in the world, Missouri has been forced to live with the metal's environmental downsides—the scars from mining, the leftover tailings ponds, and lead-contaminated soil and water. Our state bears the brunt of the damages from producing this useful but problematic metal, which persists in the environment to a troubling extent. The damage it does to the human nervous system is permanent. For the sake of our children, and their children, we must keep working to limit exposure to lead as much as possible, making sure its weighty baggage decreases through every new generation.

SOURCES

1 Robert Neuman, "Mines and Mining in Greene County," Greene County Historical Society, website: greenecountyhistorical-society.com/mines-and-mining-in-greene-county

2 Henry R. Schoolcraft, *A View of the Lead Mines*, New York, 1819, p. 48

3 Ibid., p. 96

4 Ibid., p. 120

5 Arthur Winslow, *Lead and Lead Deposits*, Missouri Geological Survey, Vol. VI, 1894

6 United States Geological Survey, "Lead Mining History," website: usgs.gov/projects/mining/historical/html

7 The Doe Run Company, "Mining and Milling," website: doerun.com/what-we-do/mining-milling/

8 *St. Louis Post Dispatch*, November 8, 1970, p. 50

9 Missouri Department of Natural Resources, "Doe Run Co. Glover Smelter," website: dnr.mo.gov/waste-recycling/sites-regulated-facilitieis-hazardous-waste-treatment-storage-disposal-tsd/mod048333875

10 *St. Louis Post Dispatch*, November 8, 1970, p. 50

11 *St. Louis Post Dispatch*, November 5, 1984, p. 91

12 *St. Louis Post Dispatch*, March 25, 2007, p. 8

13 *St. Louis Post Dispatch*, November 11, 1991, p. 102

14 *Riverfront Times*, "Doe Run Settles with EPA," October 8, 2010, website: riverfronttimes.com/news/doe-run-settles-with-epa-lead-company-to-close-herculaneum-smelter-spend-millions-2579479

15 *The Daily Journal*, July 28, 1977, p. 1

16 *St. Louis Post Dispatch*, December 9, 2017, p. 5

17 *St. Louis Post Dispatch*, March 7, 1988, p. 3

18 *St. Louis Post Dispatch*, May 10, 1983, p. 5

19 *St. Louis Post Dispatch*, May 18, 1983, p. 13

20 *St. Louis Post Dispatch*, September 27, 1989, p. 5

21 *The Current Local*, April 12, 1990, p. 1

22 *Springfield News-Leader*, June 22, 1988, p. 14

23 Interstate Technology Regulatory Council, "Oronogo-Duenweg Mining Site," website: projectsitrcweb.org/miningwaste-guidance/cs34_oronogo_duenweg.htm

24 *St. Louis Post Dispatch*, June 3, 2002, p. 8

25 Centers for Disease Control and Prevention, "Overview of Childhood Lead Poisoning Prevention," website: cdc.gov/ncah/lead/overview.html#

26 City of St. Louis, "Child Lead Poisoning," website: stlouis-mo.gov/government/departments/mayor/initiatives/resiliency/equity-youth-well-being/child-lead-poisoning.cfm#

27 Statista.com/statistics/209383/recycled-volume-of-lead-in-the-us/

28 International Lead Association, website: ila-lead.org/lead-batteries-the-most-recycled-consumer-product-in-the-us/#

29 The Doe Run Company, "Lead Battery Recycling," website; doerun.com/what-we-do/lead-battery-recycling/

30 *Springfield News-Leader*, April 30, 2008, p. 51

CHAPTER 7: HIDDEN TREASURES

"Most people are on the world, not in it—have no conscious sympathy or relationship to anything about them—undiffused, separate, and rigidly alone like marbles of polished stone, touching but separate."

JOHN MUIR

Beneath the forests, pastures, farms, and cities of southern Missouri is a landscape hidden from view, a subterranean warren of crevices, caves, and tortuous tunnels. This is karst topography, where water running on the surface often disappears into the ground, only to reappear at a spring, sometimes many miles away. Springs are the mouths, the visible parts, of large subterranean flow networks. Over time, we have learned how vulnerable these underground waters are to activities on the land surface. We have discovered that what we put on or in the ground may not stay there, but may seep into the porous subsurface. It has taken a while, but we are getting better at thinking three-dimensionally about the landscape beneath our feet.

In 1975, I began working at the Health Department in Springfield, and before long got introduced to the complexities of karst. A routine part of my job was testing water samples from private wells. While visiting one suburban property in western Greene County, I saw the well sticking up in the backyard—a 6-inch diameter steel pipe rising about a foot out of the ground. Fifty yards from it I saw a circular rock structure, about four feet in diameter, capped by a concrete slab. In the middle of the slab was a square hole covered by a hinged, steel lid. I opened the lid and stared down the hole, allowing a moment for my eyes to adjust to the darkness.

The rock structure, in fact, was an old shallow well. Across its bottom, about fifteen feet below, ran a stream of water, which disappeared into a tiny cave. This, I realized, had once been the water supply for the home. The square hole had no doubt been covered by a hand-cranked pump. But it was a plastic pipe protruding into the side of the well, about four feet below the surface, that caught my eye. From this pipe water dripped, and from the odor,

The karst topography of southern Missouri is riddled with caves and springs *(Gayle Harper)*

and visual clues, I knew this was not pure water—it was sewage from the house. The old abandoned well had been converted into a makeshift "septic tank." I had heard about people running sewage pipes "straight to the road ditch," but this was even worse—direct, malicious pollution of groundwater. But where did the water flowing across the bottom, with its load of sewage, go?

That question led to further education about karst groundwater tracing with dyes. A few weeks later, I introduced a small amount of fluorescein dye into the abandoned well. About thirty-six hours later, the dye reappeared at a spring along Clear Creek, less than a mile away. Incredibly, someone was using that spring for their home water supply. Over the years, I would learn that this kind of problem, this unhealthy connection, was common in the Ozarks, where shallow wells or spring sources were easily contaminated by septic tanks, or from runoff of cattle manure into losing streams or sinkholes, or urban runoff or gasoline or chemicals seeping from detention ponds, lagoons, pipelines, or landfills. In the porous realm of karst, the dangers are legion.

A large sinkhole formed in the bottom of the West Plains sewage lagoon *(Jim Vandike)*

One of the state's most egregious examples of groundwater pollution occurred just three years after I started work with the Health Department. Imagine a city official standing on the banks of a huge sewage lagoon, covering almost fifty acres, watching helplessly as a swirling whirlpool sucked most of his city's raw sewage, twenty-five to thirty million gallons, down into a dark, gaping hole. That is exactly what happened in May of 1978, at West Plains, Missouri, as karst topography demonstrated its cruelly capricious nature.

Geologists had warned that this could happen. In regions with karst topography, often recognized by an abundance of sinkholes, there is a constant danger of the "collapse" of ponds and lagoons. Water seeping through the bottoms of these impoundments slowly enlarges subterranean channels, as leaching water dissolves the underlying limestone. At some point, the rock is weakened to the point of catastrophic collapse, and suddenly the contents of the pond or lagoon disappear into the labyrinth below.

On April 18, 1978, just a few weeks before the catastrophic collapse at West Plains, David Hoffman, a geologic engineer with the Missouri Geological Survey Division, sent a report to DNR's Division of Environmental Quality that West Plains' sewage lagoon had a "severe" rating with respect to collapse potential. Its "integrity is in question and collapse is possible at any time," he wrote.[1] But this lagoon wasn't alone. In 1972, the Survey had compiled a detailed report suggesting that at least twenty-four sewage lagoons in fifteen southern Missouri counties posed "serious environmental threats." The lagoons served many cities, including Willow Springs, Licking, Aurora, Alton, Dixon, Mountain View, and Doniphan.[2]

The collapse at West Plains shouldn't have been all that surprising to city officials there, since the lagoon had already suffered a minor collapse in 1964, the year it was constructed.[3] In any case, the major collapse in 1978 sent public apprehension into overdrive. Within two weeks of the collapse, local health officials had reported almost 300 cases of "flu-like" illness—nausea, vomiting, diarrhea.[4] Previous dye tracing suggested the lost sewage headed southeast toward Mammoth Spring on the Missouri-Arkansas line, about 20 miles away. Geologists predicted the sewage would move at a rate of between one and three miles per day, very fast for groundwater, with scores of private water wells potentially in its path.[5]

One geologist studying the likely flow path of the polluted groundwater, Jim Vandike, had done recent dye tracing in the North Fork basin, to the west of Mammoth Spring. With his help, health officials marked off on their maps the territory likely to be affected, anticipating that many wells in the zone would need to be tested. But two weeks after the collapse, more cases of illness were reported to the west, outside the outlined area. Within a month of the collapse, more than 700 cases of gastrointestinal illness had been reported. The Army National Guard brought in two dozen tank trucks to supply rural residents with drinking water.[6]

Upwards of 40% of the wells in the zone of concern showed signs of fecal contamination. But health officials pointed out that finding fecal bacteria in a well did not necessarily mean the microbes came from the lost sewage. The culprits could also be nearby septic tanks. In fact, county and state health officials reported that even prior to the sewage lagoon collapse, an estimated 30 to 40% of wells in the region typically showed contamination, an indication that many of them were not properly constructed—that is, they were either very shallow, or lacked enough casing to prevent surface or shallow groundwater from accessing the lower, uncased part of the well.

Mammoth Spring, on the Missouri-Arkansas line, gets most of its water from Missouri *(Steve Spencer)*

On average, about 240 million gallons of water per day comes out of Mammoth Spring. It is the third largest spring in the Ozarks, after Big and Greer. This impressive size means that conduits near the spring are large, able to move huge volumes of water. Water flowing swiftly through a conduit creates low pressure around it, thus pulling water into the conduit from the surrounding flow network, rather than pushing water out into the network. This a good thing for private wells located near large conduits, because there is less likelihood of sewage in the conduit polluting the well. For the most part, Vandike suggested, the lost sewage from West Plains quickly flushed through the very open underground flow network.[7]

There were lessons to be learned all around. Rural residents needed to make sure their wells were property constructed. State agencies needed to do more thorough geologic evaluations of sites before wastewater lagoons were constructed. Lagoons are typically much cheaper to build than complicated treatment plants, but in karst areas, treatment plants are a better bet.

Meramec Spring provides water to a state-owned fish hatchery. *(Steve Spencer)*

When working properly, they produce better effluent than lagoons, and a surface discharge can be more easily monitored. The Missouri Geological Survey now performs geologic site evaluations for all proposed municipal wastewater lagoons. This should help make sure the tragic collapses of the past will not happen again.

But there are many other hazards for Missouri's groundwater. In the fall of 1981, three years after the West Plains collapse, ammonium nitrate fertilizer spilled from a pipeline break near the little town of Lake Spring. The first sign of a serious groundwater problem came when trout in the hatchery at Meramec Spring, about thirteen miles from the spill site, began dying. Dissolved oxygen levels in the spring dropped to near zero. 70,000 trout were quickly moved to another hatchery. Weakened cavefish in the spring's rise pool were rescued and sent to the St. Louis Zoo.[8] Over the next few days, the ghostly bodies of white crayfish and other cave-dwelling animals flushed into the daylight of the spring pool.

The pipeline, operated by the Williams Pipeline Company, crossed Dry Creek south of Meramec Spring. As the name implies, Dry Creek is a "losing stream," meaning that most of its water sinks into karst channels below the streambed. The leak was discovered when fertilizer, colored green to help with leak detection, stained a pool of water in the mostly dry bed of Dry Creek. Williams officials investigated and quickly repaired the leak, estimating that less than 2,000 gallons of fertilizer had leaked before the problem was fixed.[9] But when, a few days later, fish and other animals began dying at Meramec Spring, thirteen miles away, officials suspected the leak had been much larger. The seventy-five-year-old pipeline, which runs from Oklahoma diagonally across Missouri to the northeast, had leaked five times in the last five years. Two spills in Jefferson County had killed over 60,000 fish each.[10]

The lost fertilizer was easy to trace, given its distinctive chemical signature. Vandike set up equipment at the spring outlet to detect the amount of fertilizer flushing out over time. His calculations indicated that more likely 24,000 gallons of fertilizer leaked from the ruptured pipeline, twenty times the company's original estimate.[11] His study also provided details about the Meramec Spring recharge area—information that could prove useful in future emergency situations, such as chemical spills or transportation accidents.

Vandike considered it fortunate that the pipeline contaminant was water soluble, rather than an insoluble product like petroleum. The fertilizer mixed easily with water and so was flushed rapidly through the karst system. Oil and gasoline, on the other hand, do not mix with water, but float on top. Petroleum contamination in groundwater can therefore reside in pockets and crevices in the rock for long periods of time. Contaminated springs will retain their petroleum smell and sheen for many months, sometimes years, as the product very slowly breaks down or flushes out. Had the pipeline near Meramec Spring been carrying petroleum, the pollution could well have rendered the spring water unusable for trout raising for a long time.

Unlike in the karst topography of southern Missouri, groundwater in northern, western, and southeastern parts of the state moves through porous bedrock layers such as sandstones or alluvial materials like sand and gravel. Alluvial aquifers in floodplains are threatened near Kansas City and St. Louis, where, historically, much of the heavy industry located. The city of Liberty, for example, just northeast of Kansas City, draws its drinking water from alluvial wells near the Missouri River. In 1965, the city leased one of its old, abandoned drinking water treatment plants along the river to the Lee Chemical Company, which packaged and distributed industrial solvents and arranged for their disposal. Practices there must have been careless, because in September 1983, the chemical trichloroethylene (TCE) was discovered in one of the city's nearby wells.[12]

In 1983, a well at the city of Valley Park, in the St. Louis area, was found to contain TCE levels four times greater than the amount considered safe by the EPA. The well drew water from the alluvial aquifer of the Meramec River, about fifteen miles southwest of St. Louis. Wainwright Industries, a metal stamping and tool dye shop located in the river floodplain, had used TCE for degreasing parts in the 1960s and 1970s. The company ceased operation in 1979, leaving behind contaminated soil and groundwater.[13]

TCE, as it turned out, became a common groundwater contaminant all over the United States. In the book *A Civil Action*, later made into a movie starring John Travolta, author Jonathan Harr describes the tragic saga of Woburn, Massachusetts, where TCE from a chemical company fouled community wells in the 1980s. Drinking the contaminated water allegedly caused six people to die of leukemia, and many other illnesses were reported in the city. A five-month long jury trial in 1986 ended when the Grace Chemical Company was found guilty of illegally dumping hazardous chemicals on their property.[14] This was the first national recognition of the dangers of TCE contamination, and a sharp focus on the careless environmental practices of many industries.

An industrial solvent, TCE is commonly used in circuit board cleaning, dry cleaning, and parts cleaning in automotive repair. It is a dense, non-aqueous phase liquid (DNAPL). Heavier than water, DNAPLs don't mix with it, but instead sink to the bottom. These properties make DNAPLs very hard to remove from groundwater. The most common method is "pump and treat," where water is pumped from the ground and trickled over surfaces like beads or plastic plates in a columnar tank to maximize the surface area for evaporation of TCE. Air is often blown upward from the bottom of the tank to aid in this evaporation process. Even with treatment, it often takes years to rid contaminated groundwater of TCE.

A Civil Action, 1998 movie starring John Travolta *(Wikipedia)*

In 1982, TCE was found in a deep well serving the community of Republic in southwest Missouri. In the 1970s, a circuit board manufacturing facility stood about 500 feet from the city well. A fire destroyed the building in 1979. During demolition, rubble was pushed into the old basement. There were drums of TCE still stored in the basement, and an abandoned well. Concrete and bricks pushed into the hole crushed the drums, and the leaking TCE accessed groundwater through the abandoned well.[15] Before long, TCE had migrated the short distance to the city's drinking water well. By 1984, the site had become the state's number one cleanup priority, ending up on the Superfund list. Remediation of the city well lasted for many years and cost over $20 million.[16]

TCE continues to cause problems in Missouri's groundwater. In 2010, thirteen wells in Rogersville, Missouri, were found to be contaminated with TCE, six of them over the EPA drinking

An alluvial well in a river floodplain *(United States Geological Survey)*

water limit of 5 parts per billion. EPA provided filter systems for residents with polluted private wells. A source was never officially determined.[17] In 2016, Meramec Caverns had to be shut down for three months after EPA officials measured TCE at levels of concern in the air of the cave. Just a few years later, testing in the "far upstream regions" of Fantastic Caverns, a show-cave near Springfield, revealed elevated levels of TCE in cave air.[18]

The TCE in Fantastic Caverns had most likely come from the Litton Company, which once operated a circuit board manufacturing plant near the Springfield Airport. Springfield city staff alerted the Missouri Department of Natural Resources in 1982 that the waste lagoon serving the facility seemed to be leaking, as water levels were dropping. A certain amount of leakage was allowed by state rules, but seepage at the Litton lagoon seemed excessive. State attorneys filed a lawsuit, and Litton was required to drain the lagoon and spend $250,000 cleaning up the site.[19]

But problems from the lagoon weren't over. At the time the leak was discovered, the state did not require Litton to do groundwater monitoring around the lagoon, and no groundwater testing occurred on adjacent properties. By 2019, TCE contamination had shown up in private wells within a few miles of the Litton site. Of the 146 wells tested, 30 were found to contain TCE, but only two above allowable levels. By this time, Litton had been sold to Northrop Grumman, a giant company with almost $38 billion in sales in 2020.

At a highly charged public meeting in 2019, people living near the Litton Site learned more about the pollution problem. Ed Galbraith, Director of the Division of Environmental Quality for DNR, apologized to the people whose wells had become contaminated, and "for whom this came as a surprise." TCE had been detected and publicized by Fantastic Caverns in 2018, but a federal lawsuit filed in 2021 accused Litton (and therefore Northrop Grumman) of knowing since 2004 that nearby neighbors had likely been exposed. The finger pointing and lawsuits continued, but the damage had been done.[20]

The Safe Drinking Water Act, passed by Congress in 1974, required water providers to monitor the quality of their supplies and restrict pollutants and pathogens at the tap. But the act did little to keep these pollutants out of source waters in the first place. That was handled, in part, by the Clean Water Act, which addressed pollutants in wastewater discharges, urban runoff, and other major sources. But serious doubts about the effectiveness of this regulatory safety net arose when studies showed that certain chemicals could penetrate drinking water treatment systems. New Orleans, at the receiving end of sewage and industrial wastes dumped into the Mississippi River upstream, was found to have low but worrisome levels of industrial chemicals and pharmaceuticals in its tap water. What was needed to prevent this kind of problem, the EPA concluded, was more emphasis on source water protection.

I went to work in 1989 for the first non-governmental drinking water source protection group in Missouri, the Watershed Committee of the Ozarks. It had been formed five years earlier in response to concerns about development in the municipal drinking watersheds, which were primarily in Greene County outside the city of Springfield. City Utilities of Springfield, which furnished the city water, worried that county rules were not stringent enough to adequately protect water sources. The city of Springfield, the recipient of most of the public water, had no rules to govern development outside its boundaries.

A resolution to the dilemma was found in a three-way agreement between City Utilities, City of Springfield, and Greene County, creating a non-governmental advisory body. The Watershed Committee would provide public education, water quality monitoring, landowner outreach, and assistance on the development of appropriate city and county rules. It is difficult to place a financial value on pollution prevention activities, of course, but the relatively small investment in the Watershed Committee has reaped big dividends in other ways, such as increased community support for sewer rate increases; better onsite wastewater standards; stream restoration projects; and for development rules governing setbacks from sinkholes, streams, and springs. In 2024, the organization celebrated its fortieth anniversary of drinking water protection services to the community.

There are almost 3,000 public water supplies in Missouri. Most of them use groundwater. When it comes to the protection of groundwater and drinking water, prevention is key. The discovery of harmful constituents in drinking water creates anxiety and distrust of water providers. Cleanups are complicated and expensive. Lawsuits are costly, often drag on for years, and are agonizing for those involved. Clean drinking water is in short supply in many parts of the country, and the world. Missouri's clean and abundant groundwater resources aren't just important for our health and economic vitality—they are of strategic value, eyed jealously by water-deficient regions. These are hidden treasures that we cannot afford to lose.

The "old" Fulbright Spring in downtown Springfield, recently uncovered, is contaminated with petroleum products *(Loring Bullard)*

SOURCES

1 *Kansas City Times,* May 19, 1978, p. 1

2 *Springfield Leader and Press*, May 25, 1978, p. 19

3 *Daily American Republic*, May 19, 1978, p. 1

4 *Springfield Leader and Press*, May 17, 1978, p. 1

5 *Daily American Republic*, May 19, 1978, p. 1

6 *St. Louis Post Dispatch*, May 21, 1978, p. 22

7 Jim Vandike, personal interview, December 12, 2017

8 *St. Louis Post Dispatch,* December 2, 1981, p. 4

9 *Springfield News-Leader,* December 30, 1981, p. 15

10 *St. Louis Post Dispatch,* December 2, 1978, p. 4

11 Jim Vandike, personal interview, December 12, 2017

12 *Kansas City Star,* December 18, 1983, p. 79

13 *St. Louis Post Dispatch,* September 8, 1986, p. 61

14 *The Boston Globe,* July 29, 1986, p. 1

15 *Springfield News-Leader,* August 17, 1989, p. 9

16 *Springfield News-Leader,* January 9, 1995, p. 10

17 *Springfield News-Leader,* July 1, 2010, p. 7

18 *Springfield News-Leader,* January 12, 2019

19 *Springfield News-Leader,* March 16, 2019, p. 3

20 *Springfield News-Leader,* October 22, 2022, p. 3

CHAPTER 8: THE GREAT RIVER STATE

"How the river works as a maker of the landscape, sculptor, arm of creation, will always remain to some degree unknown, for it works with immeasurable leisure and patience, and often it works in turmoil."

WENDELL BERRY

The South Grand River near my hometown was, to me, anything but grand. It was more like a dirty ditch, cutting straight through the cornfields, cow pastures and pecan groves of Henry County. The banks were barren, steep, and sticky with gumbo mud; the water opaque, mocha colored. It wasn't always that way. The South Grand was once a clear, healthy prairie stream, winding through luxuriant grasslands, its gently sloping banks thickly lined with cottonwood, willow, and river birch. There were big runs of paddlefish in the spring, heading upstream to spawn.

Today, the stream is nothing like this. After devastating floods in the early twentieth century, it was totally remodeled. Making the serpentine river into an arrow-straight flume would carry water away more quickly, drainage engineers deduced, reducing flooding. With this as a goal, from 1918 to 1923 the South Grand was "channelized."[1] After this, flood flows moved downstream rapidly instead of spreading onto adjacent lands. But higher water velocities deepened the riverbed, resulting in severe head-cutting, the progression of downcutting upstream.[2] Sediment continually slumped from steeply incised banks. Undercut trees fell into the river and were swept downstream, piling up as logjams in lower reaches where the gradient flattens and the water slows.

When I went away to college, I learned that rivers in southern Missouri were very different from the South Grand. The Outing Club that I joined sponsored canoe trips on clear, beautiful Ozark streams like the North Fork, Meramec, and Big Piney. Leading the club was Dr. Oz Hawksley, perhaps best known as the author of *Missouri Ozark Waterways*, a popular float guide. Oz had a cabin at Zanoni in southern Missouri, which we used as a base camp for club floats on the nearby North Fork and Bryant Creek. Oz was a hard-nosed canoeing instructor, but he instilled in me a lifelong interest in canoeing and whitewater. More importantly, he opened my eyes to the aesthetic and ecological values of healthy rivers. In his classes, we learned that many streams had problems stemming from channelization, pollution, sedimentation, and habitat destruction.

South Grand River near Urich *(Loring Bullard)*

Nearly every river and stream in Missouri has been affected, in one way or another, by human activity. Many were altered, often unintentionally, by the effects of intensive land-uses in their watersheds—logging, mining, and farming early on, and later by development, urban runoff, and sewage. Rivers in northern, western, and southeastern Missouri were dredged and channelized in efforts to reduce flooding and drain bottomlands for farming. Hundreds of dams were built on small streams and spring branches for mill ponds, drinking water reservoirs or fishing lakes. The Missouri and the Mississippi, the state's largest rivers, were heavily altered by engineers trying to "improve" them for navigation.

The Missouri and Mississippi, however, were like brawling giants, and difficult to control. The watershed above their confluence is 700,000 square miles, an area ten times the size of Missouri. Rains and snowmelt over much of North America funnel into them, causing them to rise out of their banks, flooding farms and towns, ruining bottomland crops and sweeping away riverside businesses. In the past, the capricious rivers would suddenly shift across their floodplains during floods, leaving river-dependent communities high and dry when the water receded. In late summer and fall, flows would be reduced to braided channels running over shallow sand bars and mud flats, impeding travel by riverboats.

The Rivers and Harbors Act passed by Congress in 1899 directed the Army Corps of Engineers to eliminate "obstructions" to navigation in the Mississippi and Missouri Rivers. The Corps went about its business in typical heavy-handed fashion, straightening the rivers by slicing straight through widely looping meander bends. Major channelization efforts began in the 1920s, and by 1942, the lower Missouri River had been shortened by over 70 miles.[3] The idea was to produce higher velocities in the main channel, keeping the rivers deep through "self-scouring."[4]

Jacks Fork River, part of the Ozark National Scenic Riverways *(Loring Bullard)*

Deeper, stabilized channels allowed big boats to safely navigate the river. But with increased velocity, streambanks were subjected to higher erosive forces. The Corps fitted banks with sturdy structures called revetments to protect them from erosion. Higher levees were built to keep the rivers in flood from spreading across their floodplains. Confined to artificial channels within armored banks, their overflows straightjacketed by levees, the rivers abandoned their backwater sloughs and formerly overflowed wetlands, once prime habitats for myriad animals and plants. At the time, of course, no one gave much thought to the drastic effects these manipulations would have on the rivers' ecology.

Early in the state's history, dams were built, but mostly on smaller streams, often to power mills. The era of "big dam" building came after 1900, when the magic of electricity promised easier living, prosperity, and jobs in mining and manufacturing. *Water Power of Missouri*, a 1901 report from the University of Missouri School of Engineers, noted a wealth of potential hydropower sites in the state.[5] By the 1920s, investors and entrepreneurs had scouted dozens of potential dam sites, mostly in the Ozarks where streams had all-season flows. Here, there were relatively low populations to be displaced by reservoirs and narrow valleys where tall dams could be pinned to solid rock.

Big dams profoundly affect rivers, with flowing water ecosystems suddenly converted to still, deep water habitats. Up and downstream migration of fish and other aquatic organisms are effectively ended. Water deep in reservoirs, low in dissolved oxygen, harms aquatic life when released from the bottoms of dams; and surface water, heated in summer, stresses cool-water species when discharged over a spillway. Today, we know that reservoirs also release high levels of methane, a potent greenhouse gas, from trapped organic matter decomposing on the lake bottom.

Missouri's first big dam, constructed by the Ozark Water and Power Company, was completed in 1913 on the White River. Six-hundred feet long and fifty feet high, Powersite Dam backed water in the river for twenty-two miles. In 1922, the company announced it would next seek a license to build a dam at the "Table Rock" site upstream of Powersite.[6] Other projects were already on drawing boards by then. Kansas City businessmen promoted a hydropower project on the Osage River, and three electric companies sought rights to develop hydropower projects on the Current River. But the regulatory picture changed after 1920 with the creation of the Federal Power Commission. This powerful agency required dam developers to obtain a federal license. With license in hand, companies could proceed with complicated, expensive projects in the absence of competition.

Ozark Beach Dam at Powersite, ca. 1913 *(State Historical Society)*

Construction of Bagnell Dam on the Osage River, late 1920s *(State Historical Society)*

Bagnell Dam, completed in 1931 and bankrolled by the Union Electric Company, was the last major hydropower project in Missouri financed by private investors. The dam, 148 feet high and 2,500 feet wide, backed water in the Osage River for 129 miles. The project was seen by most as an economic boon for the region, at the time in the throes of the Depression. But there was considerable controversy over uprooting hundreds of families, to be "scattered like dust," and submerging 55,000 acres of "fertile Osage Valley farmland." Most distressingly for many valley residents, the entire town of Linn Creek had to be relocated, along with many roads and bridges.[7]

On the White River, the Ozark Water and Power Company plans originally called for a 95-foot-high dam at the Table Rock site. But the project grew impressively over the nearly four decades between inception and construction, culminating in a mammoth structure over a mile long and 252 feet high. Although at first backed by private investors, it was completed with a big influx of federal money. In 1952, Harry Truman asked Congress for $4 million to start construction.[8] By then, opposition had grown, most of it from sportsmen's clubs and conservation organizations, noting the long history of multi-day float-fishing trips on the White River.

This opposition was dwarfed, however, by the boosterism of local Chambers of Commerce, which lined up solidly in favor of the dam. The project was also supported by the political establishment. In 1953, Senator Stewart Symington said his mail was running 25 to 1 in favor of the dam. Many letters, he claimed, noted the land to be flooded was "practically useless" and "mostly ledge rock." The majority of residents, Symington insisted, wanted the dam built so they could "sell their farms in the hills and move out to better roads and better schools for their children."[9] Symington's support, especially after securing $10 million from Congress, propelled the project forward, and Missouri's largest dam was completed at Table Rock in 1958.

Environmental opposition played a much bigger role when dams were proposed on the Current River in the 1930s. The Izaak Walton League, with 165,000 members nationwide by 1927, encouraged sportsmen to oppose the dams.[10] The dams were supported by the three U.S. Representatives in the districts through which the river flows, but in the end, the magnificent scenery and clear, swiftly flowing waters won out. In 1949, both Missouri governor Forrest Smith and the Missouri Conservation Commission publicly opposed any impoundments on the Current River. Even the Chief of the Army Corps of Engineers, General Lewis A. Pick, agreed the river should remain free-flowing.

Although dams on the Current River were stopped, conservationists argued that Missouri's scenic, free-flowing rivers deserved some kind of durable protection. Several of the state's high-quality float streams ran through National Forests, and in 1958 the Forest Service stepped up its oversight, announcing special regulations on the Current and Eleven Point rivers to preserve their scenic and ecological values. But the Forest Service, primarily interested in timber production, had little experience managing recreational uses. The Park Service, in contrast, was used to handling visitors. Theodore Swem, Chief of the National Park Service, said it would be better to have some level of development and increased use on the Current River than it would be to "take a chance on its preservation without a special effort."[11]

Leo Drey and Leonard Hall—two icons of Missouri's conservation movement—entered the fray over which federal agency should provide oversight on the Current River. Drey, a St. Louis businessman, had begun buying cut over timber land near the river in 1951. By the end of the decade, he owned the largest block of privately held land in the state. Drey supported Forest Service multiple-use concepts, so naturally favored their management style. The Service had already established good relationships with private landowners in the Current River region, he noted—people who might consider signing scenic easements to protect the river.

Leonard Hall also originally favored Forest Service management. In the late 1950s, he suggested mass recreation would damage riverways, and that riverside landowners might be receptive to the scenic easement idea put forward by Drey and others.

But by the time he floated the Current River with Secretary of the Interior Stewart Udall and National Park Service Director George Herzog in 1961, Hall had changed his mind. He now felt the "multitude of self-interests" represented by watershed landowners would prevent them from "saving" the river. The locals, he suggested, almost universally practiced "destructive farming and forestry."[12]

In 1961, St. Louis Congressman Thomas Curtis introduced the "Forest Service Bill" and Congressman Richard Ichord of the 8th District put forward the "Park Service Bill." Both of Missouri's Senators, Stewart Symington and Edward Lane, supported the Park Service Bill. Their backing made a big difference. In August 1964, Lyndon Johnson signed Public Law 88-492 establishing the Ozark National Scenic Riverways (ONSR), to be under Park Service management. At the time of inception, the riverways contained 65,000 acres in the main unit and 15,000 acres of adjacent state park lands. As predicted, managing the crowds would turn out to be a monumental task. The number of canoes on the river rose from 40,000 a year in 1968 to 295,000 in 1979.[13] Since then, the Park Service has faced many more thorny management issues, including illegal vehicle access and damage from jet boats, ATVs and horse trails.

One of the state's last big dam controversies arose in 1966, when Congress appropriated funds for a dam on the Meramec River, part of a grand plan by the Corps of Engineers to build "multi-purpose" reservoirs all over the state. By 1978, the Meramec Dam had become a $167 million project.[14] But federally funded projects were by then receiving increased scrutiny. After 1970, dam developers had to submit an Environmental Impact Statement (EIS) describing, in detail, how the project would affect the environment. When the Corps released the EIS for the Meramec Dam in 1972, the Sierra Club filed suit, claiming it was too brief and far too weak.

The courts agreed. A few years later, the Corps presented a more comprehensive, 1,000-page EIS. The reaction was vigorous and largely negative. The dam would flood over 100 known caves in the Meramec Valley, including parts of two show caves, Onondaga and Cathedral. Fish and Wildlife Service officials said the project would pose a threat to several endangered species and flood 12,000 "wild woodland acres" possessing great ecological value.[15] In addition, several Corps officials had begun expressing doubts about the projected economic benefits. The views of the public had also changed. By 1978, thousands of St. Louis area citizens had floated the Meramec River or visited the area. Many if not most of them had become convinced the river should be left free-flowing.

Seeing the handwriting on the wall, Missouri Senator Tom Eagleton said he could not support the dam without a public vote, and Governor Joe Teasdale agreed. The non-binding referendum was set for August 8, 1978, with citizens in St. Louis and the twelve counties containing the Meramec Basin eligible to vote. A resounding

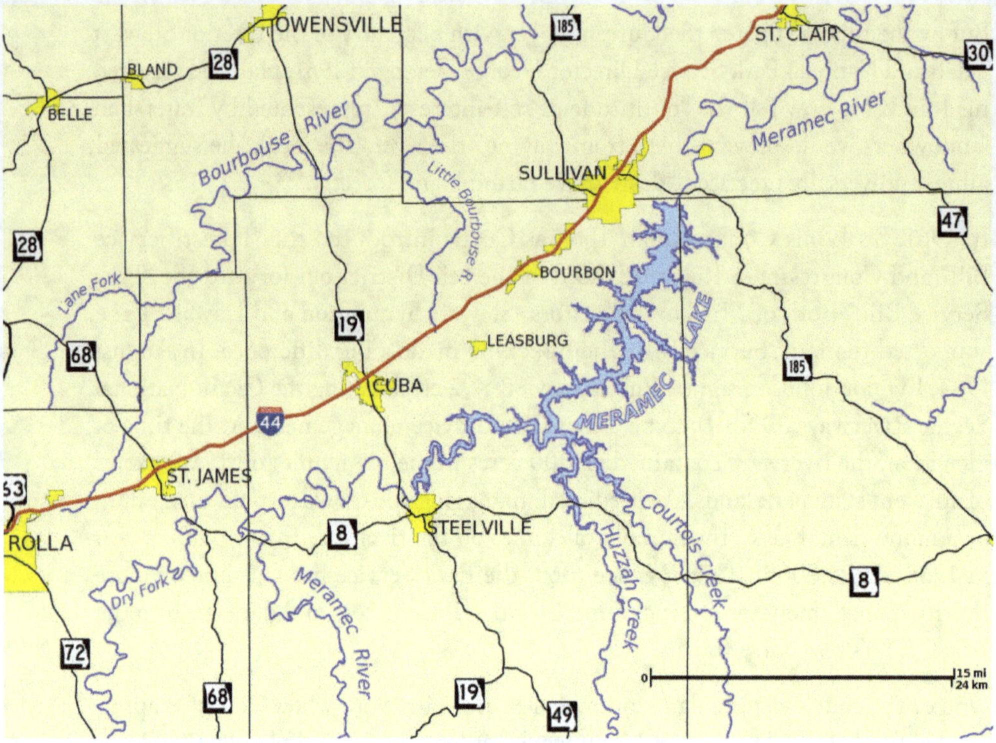

Proposed Meramec Dam near Sullivan would have flooded sections of Crawford County *(Wikipedia Commons)*

64% of voters said no to the dam, with the majority of "no" votes coming from St. Louis.[16] Although the vote was non-binding, it compelled Congress to re-evaluate the project. In December 1981, Ronald Reagan signed the papers decommissioning the Meramec Dam. Some of the already acquired lands were given to the state, doubling the size of Meramec State Park, with the remaining parcels offered for sale to prior owners. What was left went up for sale in public auctions.

By the 1980s, proposals for new dams had mostly ceased, but concerns remained that scenic, high-quality rivers were threatened by other land-uses—from farm runoff, timber cutting, streamside development, or ATVs and jet boats racing down corridors. In 1988, Leo Drey offered to commit a "six-figure sum" to finance an initiative petition for a stream protection measure, saying he was tired of waiting for the legislature to act. Ron Kucera, Assistant Director of Missouri's Department of Natural Resources, noted that streams "desperately need protection," but warned that political support for aggressive measures would be hard to find. He cited recent failures of the legislature to address the "repeated abuses of the Black River by ATVs."[17]

Drey and an environmental attorney, Lewis Green, prepared drafts of a river management proposal, the Natural Streams Act (NSA). It would direct people in the watersheds of the named "natural streams" to develop watershed management plans. These would be reviewed and approved by a five-member commission appointed by the governor. The proposal would give public agencies the power to restrict land-uses on riverbanks and charge user fees to floaters. Environmental groups, including Audubon, Sierra Club, Missouri Parks Association, and Missouri Coalition for the Environment, lined up in support, as did the St. Louis *Post Dispatch*, *The Kansas City Star*, and the Columbia *Tribune*.

Many groups actively opposed the act, however. Senator Danny Staples of Eminence noted that thirty-five organizations, including the Missouri Farm Bureau and Conservation Commission, were against it. The newly formed Missouri Landowners Association called it a "scattergun proposal" that would "create more problems than it solves."[18] The monumental vote took place on November 6, 1990. Only a simple majority was required for passage, but the measure failed, miserably, with 75% of the 1.3 million people casting votes coming out against it. Almost all the support came from larger cities, but even in St. Louis the vote was 54.5% against.

Like most states, Missouri's laws to protect its streams from pollution were weak prior to the federal Clean Water Act in 1972. But problems were recognized long before that. The state water plan in 1938 noted rising levels of river pollution, a trend also noticed by the federal government. In 1939, Franklin Roosevelt proposed a national water pollution control agency and in 1948, Harry Truman signed the nation's first water pollution law, the Water Pollution Control Act. By then, the situation in Missouri had become deplorable. The Conservation Commission noted in 1949 that 875 miles of Missouri streams were so badly polluted with sewage and industrial wastes that fish couldn't live in them.[19]

Trickling filter sewage treatment plants were used widely in the U.S. in the last century. *(Civil Images)*

Public health agencies worried about the common practice of discharging "raw," untreated sewage into rivers, potentially spreading diseases. Even large cities like St. Louis, Kansas City and St. Joseph discharged poorly treated sewage into the big rivers at their doorsteps. After 1948, with federal funding available through the Water Pollution Control Act, many cities in Missouri built their first wastewater treatment plants. The Director of Missouri's Bureau of Public Health Engineering reported in 1952 that 22 cities in the state, with a total population of over 550,000, had built "modern" treatment facilities.[20]

But river pollution in Missouri, much of it from untreated or poorly treated sewage, remained a serious problem into the 1960s. In 1965, the president of Missouri's Water Pollution Control Board pointed out that Missouri's per capita contribution to "water pollution control administration" was the lowest of any state in the U.S.[21] Even with federal grants to help cities build sewage treatment facilities, cities had to provide matching funds, which usually meant passing bond issues. In several large cities, bond issues were tough to get passed.

Over time, clean water laws, along with funding assistance, effectively ended most of the serious "point source" pollution flowing from sewage treatment plants or industrial facilities. These large and obvious sources of water pollution are now controlled through a nationwide permit system. With point source pollution mostly under control, non-point source pollution—the flow from less obvious and more diffuse sources, such as farm runoff and septic tanks—now predominates. Non-point source pollution is today responsible for over 80% of Missouri's "impaired" river miles.[22] (Impaired streams are those that do not fully meet one or more of their beneficial uses, such as drinking water, wading and swimming, or aquatic life protection).

Over the years, large amounts of public funding and technical assistance have been directed toward non-point source pollution. Cost-share programs have helped farmers install water-protective practices for crops and livestock grazing. State rules for onsite wastewater systems are more stringent than they used to be, including better site and soil evaluations. Despite these advances, non-point source pollution represents the single largest challenge today in terms of further improving the quality of the state's rivers. It is an uphill struggle. There is very little political support for regulating relatively small, widely scattered sources. And any attempts to do so are quickly condemned as government overreach. As a result, non-point source water protection efforts remain mostly voluntary. Progress has been slow and sporadic.

Recognizing that attempts to protect streams must involve people living in their watersheds, an effort began in the 1980s to establish "watershed groups." These non-governmental entities can serve as facilitators to get governments and private landowners working together to solve water pollution problems. Most watershed groups focus the bulk of their time on public education—convincing people who live, work, and play in the watershed that they must become involved in efforts to identify and address its water quality problems. But established groups often move beyond purely educational approaches, focusing energy on hands-on projects and direct outreach to landowners. Programs such as rebates for septic tank pump outs, soil testing for nutrient value, and streambank and riparian restoration projects are all becoming more common. These are important efforts, because with an increasing population and significant resistance to non-point source regulatory controls, it may become harder in the future to hold the line on the gains we've already made.

Wilsons Creek near Springfield, 1977 *(Loring Bullard)*

Boats launching on Wilsons Creek, 2022 *(Loring Bullard)*

When I moved to Springfield in 1975, Wilsons Creek, below the outlet of the city's wastewater treatment plant, looked and smelled like an open sewer. Built in 1959, the treatment plant had become overloaded. Federal funds had been made available for a plant expansion and upgrade, but construction work was still underway in 1977 when I took a picture of Wilsons Creek a few miles downstream of the plant. Water in the creek looked turbid and gray and had a foul odor.

Today, because of improved wastewater treatment and urban runoff management in Springfield and Greene County, Wilsons Creek looks a lot better than it did in 1977, when I took that picture. The city has installed successively more advanced wastewater treatment systems over the years, adding sand filtration to further clarify the treated water. An image of Wilsons Creek from the same location in 2022 shows canoes and kayaks embarking on a float trip down a sparkling clear stream.

South Grand River was channelized in the 1920s *(Jim Guldin)*

In 2023, I revisited the South Grand River, near the town where I lived during my high school years. The water looked much as I remembered it—cloudy, dun colored, flowing sluggishly. The channel, although still deeply incised, at least had more vegetation growing on its banks. According to the Missouri Department of Natural Resources, the stream remains threatened by agriculture, urbanization, channelization, and the aftermath of coal mining. But it now has a watershed group working on its behalf, the South Grand Watershed Alliance, formed in 2005. This group of citizens is committed to watershed protection "through the education of private citizens."[23]

It's a big job. The sprawling, 2,000 square mile watershed of the South Grand River covers parts of eight Missouri counties, from its "mouth" in Truman Reservoir, where it once met the Osage River, to the northwest toward Kansas City, including the cities of Clinton, Harrisonville, Peculiar and Belton. The group's website laments the fact that the watershed's "long-rooted prairie plants and forested riparian corridors" have "virtually disappeared." To help with this loss, the Alliance holds regular bank restoration and tree-planting projects. These may not be solutions to all the watershed's problems, but they are necessary parts of any solution. The South Grand may never again look like the prairie stream it once was, but through the energies of these dedicated volunteers, some of those lost values may someday be recovered.

SOURCES

1 Richard M. Duchrow, "Water Quality of the Osage Basin, 1975-1976," Missouri Projects 1-R-29, Study W-7, Missouri Department of Conservation, 1984

2 Joe Wilson, personal interview, February 8, 2023

3 William Turner, *Voices of Missouri's Rivers*, Missouri Department of Conservation, Jefferson City, Mo., 2014, p. 258

4 U.S. Army Corps of Engineers, "The U.S. Army Corps of Engineers: A Brief History," website: usace.army.mil/About/History-of-the-Corps/combat-and-military-construction/

5 University of Missouri School of Engineers, "Water Power of Missouri," Columbia, Mo. 1901

6 *Springfield News-Leader,* February 19, 1922, p. 1

7 *The Kansas City Star,* January 5, 1930, p. 32; *Pacific Transcript*, March 21, 1930, p. 4

8 *St. Louis Post Dispatch,* June 18, 1952, p. 6

9 *Moberly Monitor-Index,* March 25, 1953, p. 14

10 *Washington Citizen,* October 23, 1931, p. 1

11 Will Sarvis, "A Difficult Legacy: Creation of the Ozark National Scenic Riverway;" in *The Public Historian,* Vol. 24, No. 1, p. 38

12 Ibid., p. 44

13 Ibid., p. 49

14 *St. Louis Post Dispatch,* August 1, 1978, p. 1

15 *St. Louis Post Dispatch,* March 25, 1988, p. 1

16 Conor Watkins, Greg Hempen and J. David Rogers, "The Meramec Basin Project: A Look Back," presented to the Association of Environmental Geologists, Sept. 22, 2005.

17 *St. Louis Post Dispatch,* March 25, 1988, p. 1

18 *Houston Herald,* August 10, 1989, p. 17

19 *St. Louis Globe-Democrat,* September 11, 1949, p. 57

20 *St. Louis Post Dispatch,* March 26, 1952, p. 12

21 *Columbia Missourian,* December 16, 1965, p. 6

22 Missouri Department of Natural Resources, "Missouri Water Quality Report," website: dnr.mo.gov/document/2020-missour-integrated-water-quality-report-305b-report

23 South Grand Watershed Alliance, website: sgrwa.org

CHAPTER 9:
HELLBENDER BLUES

"One of the penalties of an ecological education is that one lives alone in a world of wounds."

ALDO LEOPOLD

When my daughters were young, I took them to the Springfield Conservation Nature Center to walk the trails and visit the exhibits. Inside the building, in one corner of the atrium, sat a big glass aquarium, and in it, lying on the bottom, immobile, was a beast that looked for all the world like a fuzzy log. It was an Ozark hellbender, a large aquatic salamander—in fact, a giant among salamanders, reaching an impressive (for a salamander) two feet long. The girls pressed their noses close to the glass. There was a single minnow swimming around in the tank, and at one point the little fish paused, seemingly unconcerned, right in front of the hellbender's nose. Suddenly, the log darted forward at lightning speed, inhaling the minnow so suddenly that my girls jumped and squealed in delight.

They, and I, have been fascinated by hellbenders ever since. They are so strange, and unique, that in their own quiet way, they demand our attention. Hellbenders live in clear Ozark streams, hiding by day under rocks, emerging in darkness to patrol for crayfish and other bottom-dwelling prey. Their habitat is cool, swiftly-flowing, highly oxygenated water. With only shrunken, vestigial lungs, they get almost all their oxygen through their baggy skin, especially the frilly folds along their sides, which are packed with oxygen-absorbing capillaries. These wavy membranes have earned hellbenders one of their nicknames: "lasagna lizards."

With tiny eyes, a blunt snout and mottled brownish color, they are not known for their beauty. One commentator suggested their ugliness is "almost unequaled" among creatures.[1] Adding to their image problems, misunderstandings about them were once common. A newspaper column in 1950 suggested the "large fellow" was "greedy, and destroys many fish."[2] A fisherman told a local reporter; "they say they can bite a kid's finger off."[3] But by the 1960s, there was recognition that finding hellbenders in a stream was, in fact, a good thing. If you see one, a biologist quipped in 1967, "you are on one of the best streams in the state."[4]

Unfortunately, Ozark hellbenders are in trouble. By the late 1980s, biologists could see that populations in some places, like the Spring River in Arkansas, were crashing. In 1989, the Missouri Department of Conservation (MDC) bestowed protection status on hellbenders, forbidding their capture or killing. Anglers were urged to release them unharmed. By 1992, they were candidates for listing under the Endangered Species Act. But questions remained about just how many remained in the wild. Dennis Figg, Endangered Species Coordinator for MDC at the time, said, "We really don't know their current status." The agency hadn't taken a statewide survey for fifteen years.[5]

By the early 2000s, MDC biologists announced there had been a 77% decline in hellbenders since the 1970s, a crash which accelerated in the 1990s. A survey in 1991 of hellbender strongholds in southern Missouri found 150 hellbenders, reporting that they seemed to be holding their own. But a statewide survey in 1998 showed almost 80% of previously identified populations had "vanished." Hardly any juveniles were found during the survey, a foreboding sign for the salamander's future.[6] A few years ago, it was estimated that less than a thousand Ozark hellbenders remained in the wild.

The drastic decline of hellbenders was hard to figure out. With further research, several potential factors in their disappearance came to light—disease, predation, over-collecting, declines in water quality. One explanation quickly rose to the top—a fungal disease that is decimating amphibian populations worldwide. Frogs have been particularly hard hit, not only in disturbed areas like cutover forests, but also in pristine, undisturbed habitats, such as high-altitude lakes and ponds. A fungus, it came to light, had quickly spread to amphibians around the world. A species of the chytrid genus, the fungus infects the amphibians' skin, interfering with its ability to uptake electrolytes.[7]

Chytrid reproduces by spores that can "swim," but that doesn't explain how it has managed to spread over several continents. Apparently, it first emanated from the highlands of South America and the eastern coast of Australia, but has since been detected in the Caribbean, Europe, and North America. Scientists still puzzle over how this rapid spread occurred, but prominent theories involve a suspiciously common vector—us. One theory posits that African clawed frogs or bullfrogs, both of which are resistant to the fungus, spread the disease when they escaped from aquariums or were purposely introduced into new areas.[8]

Another potential factor in the decline of hellbenders is predation of juveniles by an introduced species—trout. Trout, it turns out, are very fond of these tender morsels. Hellbenders are native to Missouri, living in Ozark streams for millennia. Trout, on

Ozark Hellbender *(St Louis Zoo)*

the other hand, were first introduced in the 1880s. Both like the same kind of habitat—cool, swiftly flowing streams. Scientists theorized that the evolutionary disconnect between the species meant that larval salamanders did not recognize the non-native trout as predators, so were easy prey. The stocking rate of trout in Missouri streams rose dramatically in the 1960s, exposing hellbenders to higher numbers of hungry predators.

The future of Ozark hellbenders hangs in the balance. Many professionals are hard at work to keep them from going extinct. But their status must be considered within a wider context. Animal and plant species are in trouble over the entire planet, and we are largely to blame. Scientists call the era in which we now live the "Anthropocene" because humans have "directly transformed" over half of the 50 million square miles of ice-free land on the planet.[9] With pesticides, fertilizers, fires, bulldozers, tractors, and many other instruments of disturbance and change, we have altered landscapes in all but the most inhospitable regions. In the process, we've eliminated a good portion of the planet's biological diversity without realizing, until recently, the potential jeopardy of doing so.

We have also come to realize how critical biodiversity is to our own health and well-being. Benefits provided through biodiversity are known collectively as "ecosystem services." The numbers are hard to pin down, but ecologists now estimate the value of these services worldwide to be $15 to $50 trillion annually.[10] Healthy, intact forests, wetlands and grasslands

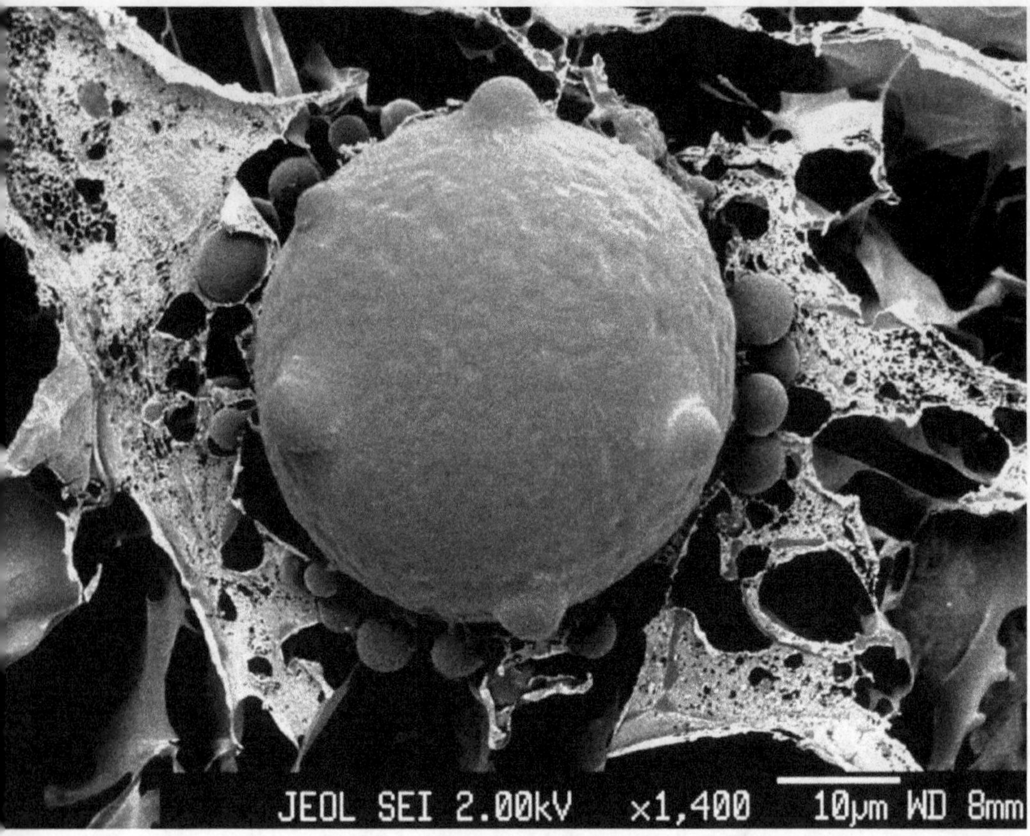

Microscopic view of chytrid fungus *(Australia Academy of Science)*

provide people with clean water and air and reduce flooding and droughts. Insect pollinators service our food crops. Bats provide U.S. agriculture a $3.7 billion annual benefit by gobbling up crop-destroying insects.[11] Important drugs and medicines come from native species, generating at least $150 billion in sales worldwide.[12] Biodiversity also enhances food security. Humans historically used over 7,000 plants and several thousand animal species for food. Today, 90% of our food comes from only about 30 crops and 14 species of livestock. Vastly simplified food production landscapes are more vulnerable to rapidly spreading diseases and disturbances such as droughts and floods.

For most of man's history, he has taken what he wanted or needed from the environment with little thought given to the long-term effects. The subsistence hunting of our ancestors eventually morphed into killing for sport and profit. Overhunting led to the extirpation, or local extinction, of several large species in Missouri. Elk, once abundant in the state, were nearly eliminated by the time of the Civil War. Black bears had nearly disappeared by 1900. Mountain lions were decimated by unregulated hunting and endless persecution by livestock owners, with the last "historic" specimen in the state killed in the Missouri bootheel in 1927.[13] Bison, although never abundant in the state, made their exit early. By 1840, they were found only in small, remnant populations.

Some game laws were in place in Missouri by the 1880s, but were rarely enforced. An editorial in *The Kansas City Star* in 1897 noted that laws intended to protect dwindling numbers of prairie chickens were "utterly ignored." Many of the farmers on whose land the birds were hunted did not object to the slaughter. What mattered most to them was that the prairie chickens were gone, since the birds were an attractive nuisance, primarily providing "amusement for the city man who comes out on Sunday to while away a day."[14] With severe depletion, the season for hunting prairie chickens was permanently closed in 1907. Today, only two small populations remain, with a total of less than 100 birds.[15]

The Lacey Act, approved by Congress in May 1900, prohibited the interstate transportation of game killed in violation of state or local laws. The biggest concern at the time was the massive slaughter of birds to supply the millinery trade. The act was difficult to enforce because it required game shippers to know state and local game laws, which varied tremendously from state to state.[16] The 1905 Walmsley Act, Missouri's first "organized" effort to protect fish and game, created a state Game and Fish Department and provided it with modest funding and personnel.[17] But the illegal hunting and shipping of game continued. A Missouri newspaper in 1908 noted that quail were being killed "in great numbers," with birds selling for $4 to $5 a dozen in Chicago and New York.[18] In 1910, 5,000 ducks a day were being "slaughtered" and shipped from Big Lake, near Blytheville in the Mississippi River floodplain near the Missouri-Arkansas border.[19]

Fishing stocks had also become severely depleted in Missouri by the 1880s. Fishing regulations were nearly non-existent. In 1895, Missouri's Fish Commissioner lambasted the "trammel-netters and dynamiters" for their extreme methods and excessive takes.[20] Amazingly, both methods were legal at the time. Attempts to bolster diminished fish populations were well intended, but typically unscientific. To some, especially local fish propagation clubs, it seemed a simple matter of re-stocking the state's streams and ponds with desirable fish. But re-stocking seldom worked because many other factors were at work in the decline of fish populations.

Early on, Missouri citizens recognized the problem of habitat loss. Prairie chickens and buffalo, for example, were dependent on extensive prairie lands for grazing and nesting, and most of Missouri's prairie lands had by 1900 been converted to crop fields and fenced pastures. The draining of land in the Bootheel early in the 20th century, mostly for agriculture, vastly reduced the habitat for ducks, geese, and wading birds. Cutting old growth bottomland forests eliminated critical habitat for the large, handsome, ivory-billed woodpecker, now considered extinct. Fish could be stocked in the state's rivers, but if river habitats had been degraded or water quality had declined, restoring populations through simple re-stocking became futile.

Damages brought by over-hunting and habitat loss were compounded by another factor—the battle against pests. Nature could deliver some devastating blows to food production, such as the "grasshopper plagues" between 1873 and 1877. In 1875, "innumerable" hoppers—actually, Rocky Mountain locusts—swept in from the western plains.[21] Nothing was safe from their gluttony. Croplands were quickly stripped bare. Chickens that gorged on hoppers could not be eaten. In response to these plagues, Missouri passed "grasshopper laws" in 1877, offering a bounty of $5 per bushel of eggs, and one dollar per bushel of "young and unfledged grasshoppers."[22] Such measures were of little help, however.

For pest control, farmers originally relied on the old standbys of sulfur, arsenic, mercury, or nicotine, but there was no way to broadcast these chemicals widely. Agricultural research in the early twentieth century led to new application methods and synthetic pesticides. One promising development was a chemical that in 1939 was found to kill a wide range of insects by attacking their nerve tissues—DDT, or dichloro-diphenyl-trichloroethane. During WWII, DDT was liberally powdered onto soldiers for lice, mosquito, and flea control, reducing cases of typhus, malaria, and dengue fever. After the war, President Harry Truman released the military's stock of DDT for use on public lands against grasshoppers, fruit flies, and boll weevils.

With government approval and support, DDT found use nearly everywhere. Health departments sprayed it onto overflowed lands to reduce mosquito complaints. Fogger trucks sprayed roadside ditches and low areas with a mixture of 8% DDT and oil, a combination deemed "harmless to pets or plants."[23] In 1951, stores sold the "Airo Kill metal DDT bomb." There was even DDT-impregnated wallpaper to control flies in kids' rooms. A researcher in 1951 claimed that of all the new pesticides being tested, "none can compare with DDT in its remarkable combined properties of high toxicity to mosquitoes, long residual effectiveness, and relatively low toxicity to plants and animals."[24]

There was no question about DDT's toxicity to mosquitoes, but the "relatively low toxicity" to other animals was increasingly debated. While powders were not readily absorbed into the skin, DDT could easily be ingested or breathed. Research showed that

DDT Posters from the late 1940s at the Roger Tory Peterson Museum in Jamestown, New York *(Loring Bullard)*

it was stored in the body's fatty deposits, especially in the liver and kidney. Concerns were raised about potential long-term health effects, including cancer. But the public paid little attention at first, even though virtually every American had some level of DDT in their body fat by the end of the 1950s. One writer suggested that DDT was so "universally used" that it had taken on the "harmless aspect of the familiar."[25]

By this time, DDT's effects on the environment were also receiving attention. Rachel Carson, an author famous in environmental circles, took on DDT in the 1940s. As a marine biologist at the Woods Hole Oceanographic Institute in Maryland, Carson paid close attention to Fish and Wildlife Service studies on the effects of pesticides on non-target organisms. She wrote an article about it for *Readers Digest* in 1945, and in a 1959 letter to the *Washington Post*, claimed that organic phosphate pesticides killed large numbers of robins and thinned the eggshells of eagles, peregrine falcons, and ospreys. Her 1962 blockbuster title, *Silent Spring*, a powerful rebuke of the chemical pesticide industry, made *The New York Times* bestseller list for six months, selling over 6 million copies. *The Kansas City Star* reported that the book raised "strong emotions all over the country."[26]

The chemical pesticide establishment did not take kindly to Carson's writings. She was ridiculed for not having a PhD or academic affiliations. American Cyanamid, a chemical pesticide manufacturer, called her book a "compendium of factual distortions."[27] Even *TIME* Magazine criticized her, mocking her findings as "hysterically overemphatic," an "emotional and inaccurate outburst."[28] But Carson stood her ground. Referring to the pesticide industry, she said, "we have put poisonous and biologically potent chemicals into the hands of persons largely or wholly ignorant of their potentialities for harm."[29]

After the debut of *Silent Spring*, many newspapers ran stories highlighting "the truth" about pesticides. "Not long ago," an editorial in *The Kansas City Times* began, the environmentally damaging nature of pesticides was something the "Audubon Society talked about, or the Izaak Walton League." Now, the question was being argued in state legislatures, scientific forums, newspaper columns, in courts, in Congress, even in the office of the president. "Why? Rachel Carson."[30] Public debate about Silent Spring eventually turned into a campaign against DDT. By 1969, it had been banned for use in some countries, and in 1972 was banned in the U.S. By the middle of the 1970s, its use had decreased significantly worldwide. But this was largely because pests had developed resistance to it, and a new generation of pesticides had been introduced.

Pesticides, of course, were only part of the problem with the harming of species. By the 1950s, conservation organizations had become alarmed by dwindling numbers of iconic species like whooping cranes—tall, majestic birds once common over much of North America—as well as bald eagles, grizzly bears, wolves, and alligators. There were less than thirty whooping cranes remaining in the wild in 1956 when a Missouri newspaper editor said, "everyone agrees that nothing should be spared" to "save them from extinction."[31] By then, a general awakening had occurred—a realization of what a world without amazing animal species would look and feel like. During National Wildlife Week in 1956, an editorial in Missouri's *Adrian Journal* quoted Aldo Leopold: "For one species to mourn the death of another is a new thing under the sun." If we can't save the bighorn sheep and condor from extinction, the writer asked, "how can we hope to save mankind?"[32]

In 1966, President Johnson signed the Endangered Species Act (ESA), allowing the Fish and Wildlife Service to spend up to $15 million per year to buy critical habitat for listed species. The Departments of Agriculture and Interior were directed to preserve habitats of endangered species on all the lands they managed. Amendments to the ESA in 1969 added a ban on the importation of wildlife under threat in any part of the world. But it wasn't long until the strong provisions of the ESA put it on the political shoals. Provisions requiring restoration plans for endangered species pitted farming or development interests against biologists pushing for habitat preservation. Legislators increasingly argued that "human factors" such as jobs and economic impacts should be given equal consideration. In 1990, Manuel Lujan, Secretary of the Interior, suggested the ESA be weakened. "Do we need to save every subspecies?" he asked. "It's just too tough an act, I think."[33]

Whooping Cranes were on the brink of extinction in the late 1950s *(Cornell Lab of Ornithology)*

The law remains controversial. Conservative lawmakers and EPA administrators have stepped-up efforts to weaken the law, backed by industry and property rights groups. Several bills that were introduced would prohibit new listings expected to cause "significant" economic harm. There have been attempts to remove gray wolves and grizzly bears from the protected list. Mike Leahy of the National Wildlife Federation said, "science is supposed to be the fundamental principle" of the listing process, but it is "getting increasingly overruled by politics."[34] Environmentalists accuse ESA officials of "slow-walking" new listings to appease critics. Species in peril must almost reach a point of no return before any actions will be taken.

The American ESA today protects about 1,700 species. Other countries have their own native flora and fauna facing extinction—their own lists. The ESA seemed like a workable strategy for saving engendered species from extinction, but it had a major flaw—it focused on saving single species, not the all-important ecosystems that sustain them. It set back many well-intentioned conservation efforts because it provided only penalties, and no incentives, for landowners who harbored threatened species on their land. Missouri, of course, is fighting its own battles to save threatened and endangered species. Our list contains 681 species classified as having the "greatest conservation need." Seventy-six of these species are listed as state endangered, with 41 of those also federally listed. Twelve species are listed as "recovering," while 17 are still "declining." In Missouri, the biggest threat has been identified as habitat loss and degradation, especially in the state's grasslands, wetlands, streams, and bottomland forests.

Kelly Rezac, Wildlife Diversity Coordinator with the Missouri Department of Conservation (MDC), noted that it's easy for a person working in her field to "get pretty discouraged." She laments that sometimes species "go down faster than you can recover them."[35] Still, she is cautiously optimistic. Trying to save species in peril is a "really good opportunity for people to work together." Although species listing and restoration occur in a highly regulatory environment, "it doesn't have to be all about regulation." Species are most effectively recovered through partnerships, she says, "by working together to do things that are good for people as well as for wildlife."

She cites as one success story the endangered Niangua darter, a small fish inhabiting clear streams in southwest Missouri. MDC worked with the Missouri Department of Transportation, Federal Emergency Management Agency, Fish and Wildlife Service and other agencies to replace low-water stream crossings with clear-span bridges, allowing fish and other organisms to migrate upstream or down. This has made a big difference for the darters, allowing them to re-occupy many sites within their historic range. Not only that, but the raised bridges are less likely to flood, so are safer for cars and school buses, as well as requiring less repair and maintenance.

Ozark Cavefish *(Missouri Department of Conservation)*

Another success story is the grotto sculpin, a white, blind fish found in caves and springs in the eastern Ozarks. In Perry County, home to many karst features, the MDC assisted a "citizen led" work force in cleaning out sinkholes, fencing cattle from springs, providing alternative water sources for livestock, and establishing vegetated buffers along streams in sculpin territory. The effort demonstrated that citizens can take a strong hand not only in protecting local species, but in improving the environment and thus, improving their own quality of life.

Aquatic species are disproportionately represented in Missouri's "declining" category. The exact reasons for decline are often poorly understood. Mussels, for example, are filter feeders living on stream bottoms. As such, they are very sensitive to water quality changes and disturbances of the stream bed, such as scouring or sedimentation. Dams and reservoirs reduce mussel habitat and separate them from their fish hosts. But there have been "mysterious" declines in some areas, where streams lose nearly their entire mussel fauna while still exhibiting a "healthy assemblage" of fish, crayfish, and aquatic insects. These problems can't be attributed to any "direct, visible cause."[36]

Another species in peril, but stabilizing, is the Ozark Cavefish, only found in about two dozen caves and shallow wells in the southwest part of the state. Biologists refer to the Ozark Cavefish as the "canary in the coal mine," the disappearance of which could indicate jeopardy for human drinking water supplies.[37] Biologists can only find and count cavefish where humans have access, so the true number of remaining fish is hard to determine. In any case, their presence, when known, can affect the trajectory of major projects. In 1978, a planned highway in northwest Arkansas, just south of the Missouri border, was rerouted two miles to avoid going over caves known to contain cavefish and gray bats.[38]

Three species of bats in Missouri are either threatened or endangered. Our state is not alone in losing its ecologically important bats. More than half the North American bat species are likely to "diminish significantly" as climate change, disease and habitat loss take their toll.[39] Millions of bats have perished from white-nose syndrome. It is thought that this fungus irritates the bat's skin, causing them to awaken from hibernation. They fly outside the cave, even in winter, looking for food. This quickly burns up the bat's energy reserves, and they grow weak with starvation. Most eventually die. A success story in Missouri, however, is the gray bat, which could soon be taken off the endangered list. Gray bats roost year-round in caves and mines, and populations have stabilized since cave gates and warning signs were installed to prevent their biggest threat to survival, human disturbance.

Worldwide, a staggering loss of biodiversity is underway. Highways, clear-cuts, cities, plantations, and croplands all divide and isolate the remaining, intact scraps of healthy forests, wetlands, and grasslands, creating "islands" of biodiversity. In isolation, with so much edge area, these biotic islands bleed species. It is difficult to predict how habitat degradation, climate change, and other stressors may act in concert to change the complexion of the biosphere. Some plants may thrive in an atmosphere containing higher levels of carbon dioxide. Some more adaptable species may be successful in wider ranges of habitats and temperatures. But many thousands or even millions of species will probably be lost. Scientists at the University of California-Berkeley have estimated that about a quarter of all species on Earth will be "committed to extinction" by 2050 under even a "moderate" climate warming scenario.[40]

Mussels, filter feeders, are extremely sensitive to disturbances of the stream bottom *(Mississippi Park Connections)*

Even as we lose biodiversity, thousands of new species are continually being uncovered in biologically-rich habitats. Biologists are racing to examine this diversity before it is lost. Most newly discovered species are small, and the majority are in tropical areas. Biologists now estimate that there are two to seven million species of tropical insects alone, in contrast with about 5,500 known species of mammals.[41] We are also finding that interactions between organisms are much more complicated than anyone previously imagined. For example, biologists are discovering how underground "mycorrhizal networks" allow communication and symbiosis among different species, a "world-wide web."[42] It is heartbreaking to lose species before we even understand how critical they may be to the healthy functioning of ecosystems.

At the same time, humans are remixing the world's biota by transporting species all over the globe. In any twenty-four-hour period, as many as ten-thousand species may be transported in the ballast water of ships.[43] Garden catalogs are filled with non-native plants; pet stores sell hundreds of varieties of non-native fish, birds, reptiles, and mammals. Non-native plants now outnumber the natives in many parts of the world. Ironically, with the artificial mixing of species, diversity has risen in many localities. Large urban areas may have thousands of species of plants, most introduced from elsewhere. The negative flip side is that overall, the number of species worldwide is declining, rapidly. The biotic world is quickly becoming homogenized, just as you see the same ten or twenty fast-food restaurants in every American city.

Parks and nature preserves help, but they will never be enough. For one thing, most of the nation's protected land is in the west, while 83% of its biodiversity lies east of the Mississippi River. And even if lands are protected, climate change may be their undoing. What appears to be a stable, protected place may eventually become inhospitable to the species living there. Most dramatically, the Anthropocene is undoing the effects of millions of years of evolution working in areas long separated by oceans, mountains, or other barriers. The breakup of Pangaea, the separating of the continental plates, isolated these areas eons ago, allowing evolution and speciation to develop millions of distinctive animals and plant species. Now, with human mixing, the biological world is becoming less complex—simpler in structure, poorer in species. Some scientists suggest we are "re-assembling the world into one enormous super-continent—the New Pangaea."[44]

The case of hellbenders provides hope for a brighter future. In April 2023, it was confirmed that zoo-raised hellbenders had reproduced successfully in the wild. It was a herculean effort on the part of resource managers. Since captive breeding and raising of hellbenders began, over 20,000 Ozark and Eastern Hellbenders have been raised and released back into the state's streams. Releasing larger juveniles into streams gives them a better chance of avoiding predators. The St. Louis Zoo, which began releasing hellbenders in 2008, was by 2021 releasing

Entomologist E. O. Wilson coined the term "biophilia" for our innate love of nature and life. *(New York Times)*

over 1,000 every year. In 2023, with its "dedicated team of hellbender keepers, life-support systems technicians, and veterinary staff," the zoo raised 1,550 larval and juvenile salamanders. MDC's Shepherd of the Hills Hatchery at Table Rock Lake hatched over 2,000.

MDC biologists have also used artificial nesting boxes with good success. Jeff Briggler, a biologist with MDC, said it takes several years for hellbenders to reach breeding age, so "capturing the event of a male attending a clutch of eggs," as happened in 2023, was "a rare event."[45] Hopefully, such events will become less rare in the future. Kelly Rezac reports that although hellbenders are still declining, the rate of decline has slowed significantly, so they are now classified as "stabilizing."[46]

How much of the Earth's biodiversity will we be able to save? The answer to that question will ultimately provide a measure of our success as a species. Will we be able to halt biological degradation and lead healthy lives surrounded by a rich, sustaining diversity of animals and plants, or will we eventually be forced to eke out an existence in a biologically impoverished world, fighting over shrinking supplies of food? We have shown how much we can damage the biosphere before understanding the perils of doing so, but we are beginning to wake up. We now have the knowledge, and the means, to make it better, if we can only find the will.

SOURCES

1 *The Daily Democrat*, September 8, 1967, p. 5

2 *St. Joseph Gazette*, April 23, 1950, p. 16

3 *St. Louis Post Dispatch*, May 29, 1973, p. 64

4 *The Daily Democrat*, September 8, 1967, p. 5

5 *St. Louis Post Dispatch*, July 7, 1992, p. 48

6 *Springfield News-Leader,* June 15, 2003, p. 8

7 Elizabeth Kolbert, *The Sixth Extinction: An Unnatural History*, New York, 2014, p. 13

8 Ibid., p. 18

9 Ibid., p. 176

10 We Conserve PA Library website: library.weconservepa.org/library-items/1043-The-Value-of-the-World-s-ecosystem-services-and-natural-capital

11 *Springfield News-Leader*, April 19, 2023, p. 9

12 Jay Withgott and Scott Brennan, *Environment: The Science Behind the Story*, San Francisco, 2011, p. 298

13 Missouri Department of Conservation, website: mdc.mo.gov/sites/default/files/1100084%20Mountain%Lions%20in%20Missour.pdf

14 *The Kansas City Star,* August 29, 1897, p. 9

15 Missouri Department of Conservation website: mdc.mo.gov/wildlife-restoration/prairie/chicken-restoration#

16 *St. Joseph Gazette*, November 21, 1900, p. 4

17 Missouri Department of Conservation website: mdc.mo.gov/magazines/conservationist/1998-09dedicate-conservation

18 *St. Louis Post Dispatch,* January 5, 1908, p. 32

19 *St. Louis Post Dispatch,* December 11, 1910, p. 38

20 Loring Bullard, "Natural Resource Protection-1904," in *Missouri Resources,* Vol. 22 No. 1, 2005, p. 5

21 *California (Mo) Democrat,* May 27, 1875, p. 2

22 *Springfield Weekly Patriot*, February 15, 1877, p. 2

23 *St. Louis Post Dispatch*, June 7, 1953, p. 135

24 *Springfield Leader and Press*, September 15, 1951, p. 19

25 *St. Louis Post Dispatch*, December 30, 1962, p. 49

26 *The Kansas City Star*, January 10, 1963, p. 28

27 Douglas Brinkley, *Silent Spring Revolution*, New York, 2022, p. 263

28 Ibid., p. 264

29 Ibid., p. 310

30 *The Kansas City Times*, March 25, 1963, p. 32

31 *Gasconade County Republican*, May 31, 1956, p. 7

32 *Adrian Journal*, March 22, 1956, p. 2

33 *The Kansas City Star*, May 12, 1990, p. 1

34 *St. Louis Post Dispatch*, August 19, 2023, p. 9

35 Kelly Rezac, personal interview, November 6, 2024

36 Missouri Department of Conservation website: mdc.mo.gov/magazines/conservationist/2017-6/missouri's-essential-freshwater-mussels

37 *The Daily Journal*, April 25, 1995, p. 32

38 *Springfield Leader and Press*, July 26, 1978, p. 37

39 *Springfield News-Leader*, April 19, 2023, p. 9

40 *The Sixth Extinction*, p. 167

41 Ibid., p. 185

42 *The Kansas City Star*, October 16, 2023, p. 8

43 Ibid., p. 188

44 Ibid., p. 208

45 *The Houston Journal*, April 27, 2023, p. 9

46 Kelly Rezac, personal interview, November 6, 2024

CHAPTER 10:
THE RISE OF BIG AG

"A healthy farm culture can be based only upon familiarity and can grow only among a people soundly established upon the land; it nourishes and safe-guards a human intelligence of the earth that no amount of technology can satisfactorily replace."

WENDELL BERRY

I have vivid memories of visiting my grandparent's farm when I was young. The farm was a few miles west of Green Ridge, a small town in Pettis County. The memories are distant now, but pieces of them are still very clear. Grandfather was a dairy farmer with about a dozen milk cows, as I recall. I remember him urging me to try my hand at milking—efforts that were pretty much wasted, of course. But it was fun to roam around the old barn, with its dim, hazy interior, festooned with cobwebs, calling to cats that were somewhat tame, but very wary. Sometimes my grandfather would give them a squirt of warm milk right from the faucet, probably as a reward for keeping the mice down. In the small, cramped kitchen of the farmhouse, there was always a pitcher of cold milk on the table at mealtime, along with a plate of fresh butter.

My grandmother took care of the chickens in a low-slung, screened-in house. I remember a few times gathering eggs with her. The dust and the smell made me a little uncomfortable. Grandma filled the feed and water containers and processed the chickens for the dinner table. I remember the dark-stained chopping block, a thick stump, where she would behead the chickens with an axe. I was horrified at the event, but in a macabre way, also drawn to it. It didn't seem to bother my grandmother at all. I figured she had done it a thousand times.

My aunt and uncle lived a few miles down the road. My uncle fed cattle on pastures and grew corn and wheat. My aunt raised their chickens in two big houses. For me, the biggest draw on their farm was the pond, about a quarter-mile west of the house. It was ringed with willow trees and very shallow, having been filled over the years with topsoil from the surrounding fields. But it was deep enough to hold fish, so I would head there with my Zebco fishing rig and a can of worms. Sometimes, my aunt would cook the bluegill I caught, or I would put them in the stock tank below the windmill for observation. When I was old enough, my dad took me quail hunting down the farm's long hedgerows.

Those memories formed my opinions of what a farm should look like—and smell like, and be like. There is still a certain satisfaction to be derived from those faded memories. My farming relatives seemed happy enough, as I recall, despite the obviously long hours of hard work. Over fifty years later, I visited other farms, this time in north Missouri, not far from the Iowa border. These were nothing like the farms of my grandparents, or my aunt and uncle, or even my sister, who with her husband still operates a beef cattle and grain farm in central Missouri. In fact, the farms in north Missouri could hardly be called "farms" at all, for they were of a more industrial nature—high technology brought to a centuries-old enterprise. The differences between my visits to Green Ridge, in the 1960s, and my visit to north Missouri, sixty years later, brought home to me the dramatically changed face of Missouri agriculture.

Farming, of course, had already changed from its early roots by the time my grandparents farmed. Prior to the Civil War, farm outputs were meager in many parts of Missouri. The First Annual Report of the state Board of Agriculture, published in 1865, said that "with the exception of some of the more important counties, very little of creditable character" could be said of the state's agriculture. But by then, the state was transitioning to a system of "improved agriculture." The Board suggested that "good farming consists chiefly in deep plowing, under-draining and manuring."[1] The state's farmers increasingly adhered to these basic maxims until newer, more "improved" methods came along.

In the early days, farms were very diverse, raising animals and plants together. Animals not only provided labor and meat, their fecal wastes became fertilizer for crops. Farms produced manageable amounts of manure and thus, didn't require commercial fertilizers. The "genius of American farm experts," Wendell Berry suggests in *The Unsettling of America*, is "they can take a solution and divide it neatly into two problems."[2] When tractors did the work, animals weren't needed, so there was no manure for fertilizer. A sole focus on crop production, or on livestock raising, negated their complimentary values.

After the dust bowl in the 1930s, farmers focused their efforts on providing protection against wind and water erosion, including windbreaks and soil-protecting cover crops between harvests. Cover crops were also important for disrupting the development of insect pests. No one crop was on the land long enough to attract a big infestation. But things changed significantly in the middle of the twentieth century, with the introduction of synthetic fertilizers and broad-spectrum pesticides.

After WWII, President Truman urged farmers to "conglomerate and automate."[3] Richard Nixon's Secretary of Agriculture, Earl Butz, instructed them to plant "fencerow to fencerow" in order to maximize production. This meant more reliance on pesticides and herbicides, and eventually, the use of genetically engineered crops. With the adoption of these new methods and technologies, crop and livestock yields shot up impressively. But there was a catch, as many hard-working farmers eventually realized. As they became more efficient, they began over-producing, resulting in lower prices for their products. More and more, in the 1960s and beyond, farmers had to raise more crops and livestock just to stay afloat.

Along with escalating production came ever more severe environmental problems. The "Corn Belt," which includes northern Missouri, was prairie land when the first settlers arrived in the mid-1800s, with up to sixteen inches of fertile topsoil. At least half of this topsoil has eroded away since the first plowing. Corn and soybeans came to dominate plantings in the Corn Belt, so that today the landscape is nearly covered by just these two crops. NASA reports that corn and soybeans are grown so tightly in the Corn Belt that in July, the region generates more photosynthetic activity than any other place on Earth.[4] Today, the almost total dependence on the corn-soybean rotation is "quietly exhausting" a prime resource—topsoil—because land is barren and exposed between the corn and soybean rotations.[5] Tom Philpott, author of *Perilous Bounty*, warns that the Corn Belt has emerged as a "machine for sacrificing soil."[6]

Much of the eroded sediment, pesticides and fertilizers find their way to the Gulf of Mexico. The plant nutrients in fertilizer, especially nitrogen and phosphorus, contribute to widespread algae blooms near the outlet of the Mississippi River. When algae die, they are consumed by bacteria, which use up oxygen in the process. This leads to an anoxic, or low oxygen area, in the gulf, a "Dead Zone," now about 7,000 square miles in extent, where fish and shellfish can no longer live. Midwestern states like Missouri contribute a disproportionate share of nutrients to the zone. Our state alone is responsible for 17,000 tons per year of phosphorus and 144,000 tons of nitrogen.[7]

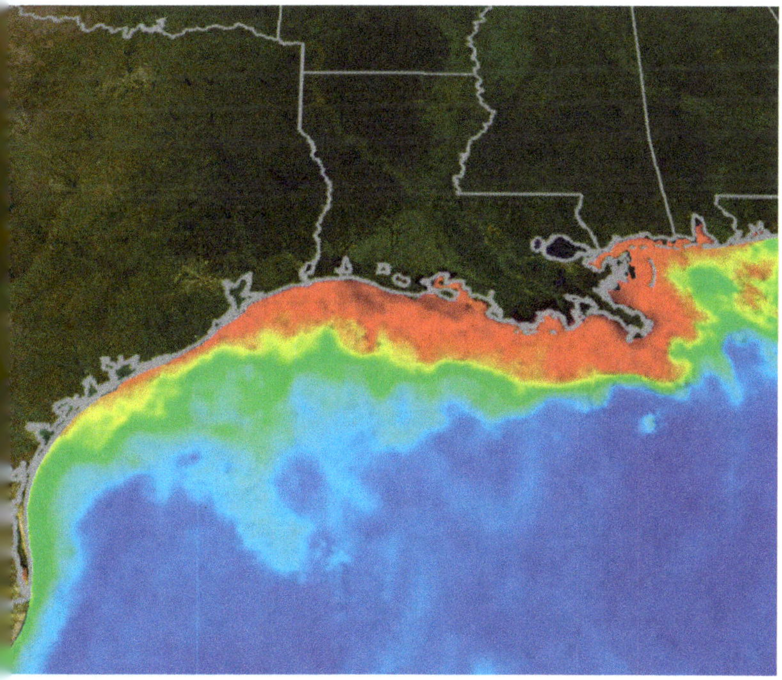

Low oxygen "dead zone" (in red) now spreads over 7,000 square miles of the gulf coast *(National Geographic Education)*

Farmers in the Midwest have largely become dependent on pesticides and herbicides for high yields. The Monsanto Company, based in St. Louis, rolled out its first glyphosate-based herbicide in 1974, calling it Roundup because it "surrounded" and killed most kinds of weeds by interfering with their enzymes, keeping their roots from absorbing nutrients. Monsanto also led in genetic engineering, using gene transfer technologies to produce genetically-modified organisms, or GMOs. By the mid-1990s, the United States Department of Agriculture (USDA) had greenlighted a brand-new Monsanto GMO product, the "Roundup-Ready" seed, which produced a plant resistant to the company's flagship herbicide.

But problems developed when weeds began to develop a resistance to Roundup. The first pesky weed to show this resistance was water hemp, which appeared in a Missouri soybean field. By 2007, more than half of the counties in northern Missouri had sprouted these new super-weeds. In anticipation of more widespread resistance developing, Monsanto and other companies began engineering seeds resistant to more than one herbicide. Monsanto, for example, claimed its dicamba-Roundup "cocktail," introduced in 2015, would make it difficult for weeds to survive two herbicides attacking simultaneously in two entirely different ways.

But David Mortensen, an Iowa State University weed specialist, predicted the chances were "quite high" that these products would unleash a "new generation of superweeds" resistant to both herbicides. Farmers, caught in the middle and fighting losing battles with weeds, typically responded by applying even heavier doses of herbicides. They had become locked into a "permanent chemical war against ever-evolving weeds," with "Monsanto and its peers playing the role of defense contractors."[8]

And there were other problems. By the late 2000s, the USGS had found traces of glyphosate in samples of air, rain, and streams. There were concerns that there could be widespread human exposure and negative health effects. In 2015, the World Health Organization listed glyphosate as a "probable human carcinogen," a charge Monsanto vigorously denied.[9] 2-4-D, another commonly used weed killer, had already been found to be linked to endocrine gland disruptions. Dicamba was receiving attention because the herbicide was easily volatilized, resulting in the problem of airborne "drift." Pesticide drift could kill crops in nearby fields that were not herbicide-resistant, and other plants in nearby orchards, vineyards, and gardens.

Lawsuits over the potential for glyphosates to cause cancer, especially non-Hodgkin lymphoma, proliferated after 2000. By 2024, Bayer, which bought Monsanto in 2018, had been hit with 170,000 lawsuits. The company set aside $16 billion to settle the cases, but looked to the state of Missouri, where many of the lawsuits originated, to shield it against further lawsuits and limit damages that could be collected.[10] Legislators in Iowa, Idaho and Missouri introduced bills limiting liabilities for agribusinesses. In Missouri a bill was introduced in 2024 but failed to make it to the full session. All this legal strife has hurt the bottom line for Bayer, which in the first quarter of 2024 shed over 1,500 employees, or about 1 ½ percent of its worldwide workforce.[11]

As in other states, animal agriculture in Missouri became more industrialized in the twentieth century. There had long been chicken houses in the state, but the numbers and sizes of poultry operations increased exponentially after the 1970s. The so-called "vertical integration" of the poultry business, where one company owns an assembly-line process including hatcheries, growing houses, feed mills and meat processing plants, has been credited to John Tyson. In the early 1930s, Tyson was preparing to inherit his parent's farm outside Kansas City. But Depression-era farm income was barely enough to keep his parents afloat, so John took the old truck and with his wife headed south, to Arkansas, ending up in the little town of Springdale. He soon found his way into the poultry business, which was starting to boom.

Tyson made good money hauling chickens to the cities but soon realized there were also profits to be made in selling chicken feed. He started a feed mill and provided feed to growers up front, collecting from them when the chickens were sold. Before long, Tyson had invested in an interconnected network of hatcheries, feed mills, and growing houses. In the 1950s, the company began contracting with growers to raise birds supplied by Tyson. Growers owned the flocks, so were more motivated to take care of them—staying up late, paying attention to feed bins and watering troughs and inside air temperatures. It was better for farmers, at least initially, because they knew the price they would receive for the birds, reducing the risk from selling in a wholesale market.

Contract chicken grower near Seligman, 1958 *(State Historical Society)*

There was no question that the model developed by Tyson produced meat efficiently and provided companies with steadily increasing profits. Other big companies, raising poultry or pork or beef, eventually adopted the Tyson model. These new industrialized meat operations patiently worked out all barriers to their efficiency and expansion into new areas. For example, Premium Standard Farms, a nationwide pork producer, looked to expand its hog operations into northern Missouri in the early 1990s. The company had plans to raise about two million hogs in three Missouri counties near the Iowa border. But many people, including legislators, were wary of corporate-controlled agriculture. A law passed in Missouri in 1975 banned large corporations from owning and operating industrial farms.

Premium Standard figured out a way around this barrier. The company used hefty portions of its profits to support political candidates who would look out for the company's interests. In the final hours of the Missouri legislature in 1993, an amendment was tacked onto an economic development bill exempting the three counties in North Missouri—Sullivan, Putnam and Mercer—from the corporate farming law. Farmers in Putnam County fought back against corporate agriculture. Lincoln Township passed a zoning ordinance mandating setback distances between industrial hog farms and residences and requiring a security bond on waste lagoons in the event of failure. But one day after the ordinance passed, Missouri DNR issued letters of approval to Premium Standard for the construction of nine lagoons.

Claiming the zoning ordinance would constitute an illegal "taking" of private property, Premium Standard sued Lincoln Township and demanded $7.9 million from it to cover the company's investments. The township's cause was supported by Willie Nelson, organizer of "Farm Aid" in 1985, who attended a rally there with about 3,000 opponents of corporate hog expansion. But in May 1996, a Putnam County District Court ruled that the law giving zoning power to townships was unconstitutional. The township appealed the ruling to the Missouri Supreme Court, which confirmed that agricultural operations were, in fact, exempt from zoning laws.

Hog raising in 1959 *(State Historical Society)*

Industrial-scale hog operations proliferated in north Missouri. "Spills" of hog wastes were a perennial problem. Most of the operations used the same methods for capture and treatment of wastes. Hog houses have slatted floors, allowing wastes to drop into deep concrete pits. From here, the liquid waste is pumped into outside lagoons, which may or may not be covered to trap methane and reduce odors. Large farms commonly have many acres of adjacent land where they can "land apply" wastes from the lagoons, which is intended to provide fertilizer value for pastures or crops. The companies are required by the Missouri DNR to make sure the wastes stay on their land and don't run off into adjacent creeks or lakes. This, of course, is easier said than done.

In the event of a leaking or clogged waste line, or any number of other problems, the company is to provide "emergency containment" for any wastes spilled on the ground. The containment structures are often small earthen berms thrown across a drainage to contain the wastes, allowing them to infiltrate into the soil. But if it happens to be raining during the emergency, or if the containment structure is overwhelmed, wastes can make it to the closest waterbody, causing, in some cases, significant contamination with fecal bacteria and high-organic content waste materials. The decomposition of organic wastes robs streams of oxygen, often causing fish kills.

Unfortunately, spills, which must be reported to DNR, became routine. In 1995 alone, Premium Standard reported seven spills from operations in north Missouri, killing about 173,000 fish.[12] At the Wolf-Brantley Complex, a cluster of Smithfield hog houses near Princeton in Mercer County, DNR recorded 164 separate spills between 1991 and 2021.[13] The average volume of waste per spill was just short of 6,000 gallons, with an average of 26,000 gallons spilled per year. Regulators became even more concerned about such spills after 1997, when runoff from hog operations in the Chesapeake Bay watershed contributed to an outbreak of pfiesteria, a toxin-producing organism that feeds on algae.

Odors from hog lagoons and application sites have also caused problems. Companies buy land around growing houses not only for waste application, but also to act as buffers against neighboring properties. After odor complaints and lawsuits from nearby residents in the late 1990s, Premium Standard at first threatened to leave the state, but eventually agreed to deal with the problems. In a 1999 consent decree, Missouri Attorney General Jay Nixon forced the company to spend $25 million to reduce air and water pollution. The company tried covers for lagoons, scrapers to remove waste from pits, and biofilters, but failed to come up with good solutions. The complaints continued. In 2020, Premium Standard was sued by 300 people over odor nuisances, and a jury awarded the plaintiffs $11 million.[14]

The issue of industrial hog facilities tends to divide communities, even in rural parts of the state. Those who work for the company or sold land to it say they are happy; neighbors living near the facilities are usually not. Some say the hog facilities smell okay; others say the

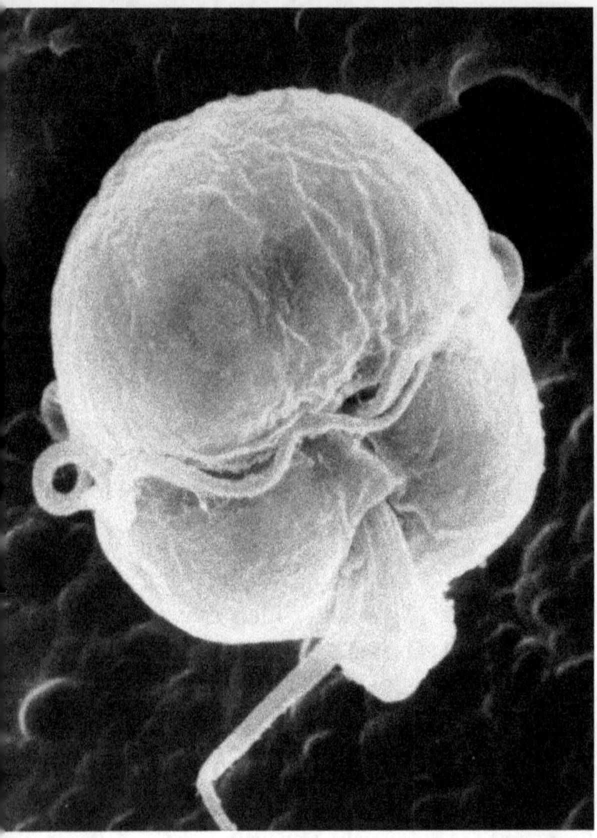

Pfiesteria, a parasitic dinoflagellate, caused large fish kills in the Chesapeake Bay region *(Wikipedia)*

odors make them sick. Fast-food and convenience stores benefit from their presence, but farm-related businesses suffer. According to Marvin McDonald, a rancher in Mercer County, the benefits from hog operations didn't "offset damage to the area's air, water and quality of life." Bob Whitare, president of Farmers Bank in Sullivan County, suggested that local businesspeople were afraid to criticize the company, for fear the area's biggest employer might pull out.[15]

Big Ag has been working for decades to hedge their bets against government interference. Smithfield and other companies spend hundreds of thousands of dollars attempting to influence politics in Missouri. They also get deeply involved in state regulatory and rule-making processes. Water pollution rules are developed and enforced by the Clean Water Commission. When it was established, the Commission was to have at least four of the seven members represent the "general public," while no more than two could represent agriculture or mining. In the waning hours of the 2016 legislative session, on a voice vote after only a few minutes of discussion, an amendment was added to an unrelated bill that would require *no more than* four members of the general public to sit on the commission, while *at least* two members had to be from mining or agriculture.

The amendment was introduced by Brian Munzlinger, who had received a large campaign contribution from Smithfield. Ray McCarty, president of Associated Industries, defended the move, saying that "just because someone has experience in an industry doesn't mean they're going to side with the industry." McCarty also suggested Commission members should have "expertise in the field they are trying to regulate."[16] Governor Jay Nixon vetoed the bill, charging that the proposed change had never been publicly vetted. Two months later, the legislature overrode his veto. Within two years, Governor Eric Greitens had appointed three new members to the Commission, all with strong ties to the industrialized farming lobby.

After the 1990s, some of Missouri's counties and townships began to follow the lead of Lincoln Township—exploring legal options for preventing, or regulating, industrial-scale CAFOs. By the late 2000s, about twenty counties had passed "health ordinances" intended to limit industrial operations through setback distances and other means. In 2019, Senator Mike Bernskoetter sponsored legislation intended, he said, to level the playing field for agribusiness in response to what he called a "patchwork of arbitrary and highly restrictive county ordinances written by non-experts." Proponents of Senate Bill 391, meant to supersede local control, said the current "hodgepodge" of regulations could stall the expansion of "modern agriculture techniques in the state."[17] The Missouri Farm Bureau, Missouri Cattlemen's Association, Missouri Corn Growers and other groups allied with agribusiness supported the bill.

But according to the Missouri Rural Crisis Center, "tens of thousands" of Missourians opposed it. The fact that it passed on a party-line vote of 103 to 44 demonstrated, to the Crisis Center, "the immense disconnect between the will of the people and their elected representatives."[18] Governor Parsons called the passage of SB 391 a "big win for Missouri farmers, ranchers and agribusiness," a move that would allow Missouri to "lead the way in meeting a growing world food demand."[19] More than anything else, however, it showed the stranglehold Big Ag had on the state's legislature.

Industrial meat companies strive for the greatest efficiency possible. The bottom line is profit, as much or more than "feeding the world." In fact, Missouri agriculture does little to feed the world. Eighty-five percent of U.S. food exports go to twenty highly developed nations, which have few "dire" hunger problems. The few dozen countries with the biggest hunger problems get very little food from us.[20] Furthermore, any little dip in profits can cause companies to close portions of their processing chains, temporarily or for good. Communities dependent on these industries for jobs are suddenly left holding the bag. This happened in Noel, Missouri in the fall of 2023, when the Tyson chicken processing plant closed, throwing 1,500 people out of work.[21]

Big Ag is fond of portraying the "battle" over agriculture as pitting urban folks, who just don't understand how agriculture works, against "family farms." This is misleading, but the threat of interference has served to mobilize many in the Ag community against what they see as outside, largely urban, agitation. Missouri's agribusinesses sought to involve counties in their "fight" to protect agriculture when they launched Missouri Farmers Care (MFC) in 2015, supported by a coalition of 45 agribusiness groups. The stated purpose of MFC is to support "agricultural stewardship, growth, and opportunities," but an important corollary is to enroll "Agri-Ready" counties—those agreeing to have no ordinances pertaining to agribusiness any stricter than state law. As of 2024, about two-thirds of Missouri's 114 counties, almost all in rural areas, have enrolled.[22]

Agriculture contributes almost $100 billion annually to the state's economy and provides a half-million jobs. But agriculture has strayed far from the bucolic image of the family farm lodged so pleasantly in our minds. Industrial agriculture, recognizing the symbolic power of that image, has co-opted it, making "save the family farm" their rallying cry. The implications for the environment are immense. Today's industrial agricultural enterprises are "consuming the ecological foundations that support agriculture itself," through losses of healthy soil, nutrient and pesticide pollution, and declining reserves of groundwater.[23] Richard Manning,

Logo for the Agri-Ready Program, Missouri Farmers Care *(Missouri Farmers Care)*

in *Against the Grain*, suggests that the Dead Zone in the Gulf of Mexico, blotting out life in what was formerly one of the continent's most productive fishing regions, means a "high quality source of low-cost protein is being sacrificed so that a source of low-quality, high-input subsidized protein can blanket the Midwest." Big Ag profits, he claims, largely by "squeezing farmers" and "offloading the costs of the ecological degradation they cause onto communities and the taxpayer."[24]

Big Ag does not necessarily represent the future of agriculture. In fact, it offers a poor choice for a sustainable future. Missouri, like most agricultural states, needs to make big investments in regenerative agriculture—strategies that will build healthy soils, conserve water, and restore biodiversity, all while allowing farmers to make a decent living. Rob Myers, Director of Missouri's Center for Regenerative Agriculture, says regenerative practices are catching on in the state, especially cover crops, which are planted between rotations of the cash crops like corn and soybeans. Cover crops are now used on about a million acres of Missouri cropland, ten percent of the total cropped acres.[25]

Cover crops, such as rye, clover, hairy vetch, radishes, and turnips not only reduce soil erosion, but also help control weeds. As cover crops grow, die off and decay, they provide food for beneficial soil microbes and create air pockets to aid water infiltration. Growing a wider variety of crops requires more labor and management but is balanced out by higher productivity and reduced expenditures on agrichemicals. There is also more interest these days in returning biodiversity to farms—not only multiple crops, but putting animals and plants on the same land, "a little like the buffalo on the prairies," Myers suggests. Cover crops like cowpeas, millet and sorghum, often planted together, can be grazed part of the year. Grazers, and their wastes, improve the diversity of biological life in the soil, making it even better for growing plants.

In March of 2024, I traveled to north Missouri to see for myself the settings of industrial hog farms. I took pictures, but "biosecurity" signs warned me to stay a considerable distance away. I could see and hear gigantic fans at the ends of buildings working to keep the hogs cool. I was told that if the fans quit on a hot day, pigs could start dying within hours. Near one row of buildings was a large lagoon, probably an acre or more in size. Riser pipes protruding from the surrounding fields provided attachment points for irrigation systems where wastewater could be sprayed onto the land. On the day I visited, there was little detectable odor, but no wastes were being sprayed at the time.

I visited the hog farms with Doug Doughty, a fifth-generation farmer in Livingston County. Doug and his wife own 1,200 acres where they grow livestock, corn, wheat, and soybeans. They use many of the conservation practices recommended by the soil and water district, including vegetated buffers around fields and native prairie strips seeded with wildflowers. After our visit to the hog farms, Doug and I had breakfast in Chillicothe, the largest city around. I asked him why he was concerned about industrial-scale hog farms. "Look at our neighbor to the north," he said, referring to Iowa, two counties away. "Half of their rivers and streams are impaired." He blames the problems on CAFOs. "It's obvious when you look at those watersheds and the amount of manure that's produced." Doug is not anti-CAFO, but he is concerned that their proliferation could "overwhelm the environment with nutrients and air pollution."

Large hog houses near Princton in Mercer County *(Loring Bullard)*

It's not just the manure. Many farmers apply both manure and commercial fertilizer, using manure as a back-up plan, an "insurance policy," and end up applying nutrients well beyond the agronomic needs of their crops. Doug notes that statistics from Iowa show commercial fertilizer sales have remained level, while the amount of CAFO manure being spread has increased. "If nutrient management plans were being strictly followed, wouldn't you expect fertilizer sales to decrease?" The nutrient composition in CAFO pits and lagoons can be variable and "inconsistent," he suggests, while commercial fertilizer analyses are precise and consistent. Farmers often end up applying the "regular" amount of commercial fertilizer to make sure they have enough nutrients getting to the crops.

Clusters of Smithfield hog houses in Mercer County *(Google Earth Image)*

Doug is a highly educated farmer and follows the science of nutrient management. He acknowledges the Dead Zone problem and suggests farmers carefully consider the "nutrient processing" on their fields. He uses a comparison with livestock. "We are focused on containing our livestock; if they get out, there is panic, a sense of urgency," because they're valuable and "could endanger other people or their property." But farmers have a "nonchalant" attitude about their nutrient applications, and "what happens when there is a big rain." Doug wants farmers to have the same attitude about nutrients as they have toward their livestock; "they're valuable, and can be a liability to other people, so let's do a better job of containing them."

There are many practices to help them do this, such as buffers around fields to reduce runoff, conservation tillage, or even applying less fertilizer. "It's not that every farmer should be doing all of these practices, but every farmer should be doing something." Doug mentioned road ditches, of which there are thousands, potentially serving as linear, vegetated wetlands to capture and reduce nutrients, rather than grading ditches down to "gravel and dirt." From his reading about research by Dr. Keith Schilling of the Iowa Geological Survey, he suggests this system of vegetation-enhanced drainageways could have major impacts on the quality of water reaching rivers and lakes.[26]

Even more recently, I made a trip back to see the farms of my childhood, my relative's places west of Green Ridge. My grandparents' house is still there, although extensively remodeled. The acreage behind the house, where my grandfather once grazed milk cows, backs up to the Katy Trail. While cycling on the trail, I stopped on a high trestle to view the old farm from an entirely different perspective. The old barn is gone, as is the weedy hedgerow that once defined the farm's eastern boundary. Instead, there are clean fencerows and crops growing where little patches of forest once stood. Instead of a creek under the trestle, there was a straight muddy ditch. I thought about my time long ago at the creek, under the trestle, wading and catching crawdads; looking up as diesel locomotives thundered overhead. Times have changed, all right. I remember that as I stood atop that trestle, I felt an aching in the pit of my stomach. Somehow, I thought, something precious had been lost.

SOURCES

1 First Annual Report of the Missouri State Board of Agriculture, Jefferson City, 1865

2 Wendell Berry, *The Unsettling of America: Culture and Agriculture*, Berkeley, 1977, p. 66

3 Douglas Brinkley, *Silent Spring Revolution*, New York, 2022, p. 27

4 Nasa.gov/news-release/satellite-shows-high-productivity-from-u-s-corn-belt/

5 Tom Philpott, *Perilous Bounty*, New York, 2020, p. 8

6 Ibid., p. 7

7 KCUR.org/news/2023-02-105/

8 *Perilous Bounty*, p. 113

9 *St. Louis Post Dispatch*, August 11, 2018, p. A8

10 *St. Louis Post Dispatch*, May 23, 2024, p. A6

11 Ibid., p. A6

12 *The Kansas City Star*, December 28, 1995, p. 24

13 Missouri Department of Natural Resources, *Spill Characteristics and DNR Enforcement Efforts at Badger/Wolf/Brantley CAFO Complex*, 1991-2021

14 *Columbia Daily Tribune*, August 31, 2012, p. 17

15 *St. Louis Post Dispatch*, August 18, 1996, p. 58

16 *The Kansas City Star*, June 10, 2016, p. 1A

17 *Columbia Daily Tribune*, May 15, 2019, p. 6

18 *Columbia Daily Tribune*, May 15, 2019, p. 1

19 *Douglas County Herald*, June 6, 2019, p. 6

20 *Perilous Bounty*, p. 94

21 *Springfield News-Leader*, October 8, 2023, p. 1

22 Missouri Farmers Care website, mofamerscare.com

23 *Perilous Bounty*, p. 5

24 Richard Manning, *Against the Grain: How Agriculture has Hijacked Civilization*, New York, 2004, p. 100

25 Rob Myers, personal interview, November 1, 2024

26 Doug Doughty, personal interview, March 21, 2024

CHAPTER 11: AN UNPARALLELED EXPERIMENT

"Like music and art, love of nature is a common language that can transcend political or social boundaries."

JIMMY CARTER

On the 12th of May, 2017, I stood staring at a scene of utter destruction. Before me, a highway that once crossed a deep valley flew off into oblivion. Just a few days before, a concrete bridge had stood there, crossing high above the North Fork River. Melissa Burnett, a resident of the village of Dora, about five miles west, had seen the bridge shortly before its demise—had seen a tangled mass of trees lodged in it. The juggernaut of rushing water and the tremendous leverage of wedged trees finally tore the 300-foot-long concrete bridge off its piers and brushed it aside like a tinker-toy, its twisted wreckage coming to rest on the bank several hundred yards downstream. The engineers who designed the bridge in 1975 most likely believed this would never happen—their sleek, modern structure folding like a house of cards.

It wasn't just the bridge. The channel of the North Fork River, which formerly looped around gravel bars and paused in deep holes, resembled a giant, arrow-straight flume after the flood. Both sides of the flume were lined with fallen trees, all pointing downstream, their bases buried in huge mounds of gravel. A little farther downstream, at the Sunburst Ranch, the main cabin had been swept off its foundation, along with the camp store, barn, and three smaller cabins. After analyzing rainfall data, the National Weather Service determined that the flood had been a thousand-year event, meaning that it had a one-tenth of one percent probability of happening in any given year. It was a toad strangler, dropping an estimated ten to twelve inches of rain in the North Fork watershed upstream of the bridge.

One-tenth of one percent—those are extremely low odds; and yet, it happened. In fact, it's happening in rivers all over the Midwest. At the James River in southwest Missouri, a stream gage on the lower river has been measuring flows since 1921, over a hundred years. As of 2024, four of the five highest flows ever recorded were within the last sixteen years, and all five were in the last 31 years.[1] The probability of certain flooding events is based, of necessity, on historic data. But things have changed since the first measurements were taken. In the James River, for example, there have been significant alterations in land-use and vegetative cover. There are now many more acres of impervious surfaces like rooftops, roads and parking lots that send runoff downstream rapidly. But those effects are most likely minor compared to the main agent of change—more intense rainfall occurring ever more frequently. The weather is changing.

Weather is the day-to-day changes in rainfall, temperature, and humidity, while climate is the longer-term trends in weather patterns. Either way you look at it, the changes are real. And it's not just happening in Missouri, of course. As I write this, hurricane Beryl is hurtling toward the Yucatan Peninsula of Mexico. Beryl has the distinction of being the first ever Category 5 hurricane to form in the Atlantic Ocean prior to July 16. It has already wreaked havoc on Jamaica, an ever more common occurrence in the islands of the Caribbean. Wildfires are already burning in California, and will no doubt soon plague the Rocky Mountains, where the fire season has been lengthened by at least a month since 1979.[2]

There can be little doubt that climate change is upon us, any more than we can deny that carbon dioxide levels in the atmosphere, a primary driver of climate change, are reaching their highest levels since measurements began. 2023 global emissions of carbon dioxide reached almost 37 billion tons, a new record, 1.4 % above the pre-COVID level. Some parts of the

Hammond Bridge over the North Fork River was swept away in the flood of 2017 *(Springfield News Leader)*

world are seriously focusing efforts to curb greenhouse gases. Others are lagging. According to the Global Climate Budget website, carbon dioxide emissions are currently declining in 26 countries, but these efforts will be "insufficient to reverse the growth in global fossil fuel emissions.[3]

There is also little doubt that humans are mostly to blame for climate change. It is difficult to determine precisely how much is human-induced, and how much is "natural" variability, but all the evidence suggests our increased loading of carbon dioxide has accelerated the greenhouse effect, the trapping of heat near the earth. A warm atmosphere allows it to hold more water vapor. In the case of Missouri, warm ocean waters in the Gulf of Mexico evaporate mega-tons of water vapor into the warm atmosphere, and prevailing winds carry it northward before cooling and dumping it over the Midwest, often in very intense bouts of rainfall.

Intense rainfall causes creeks to swell rapidly. Creeks converge into rivers, and rivers rise out of their banks. Rivers in flood do a lot of work, which translates, for us, into lots of damage. In Ozark streams like the North Fork and James River, riverbeds consist primarily of gravel. Floods pick up and move this gravel in massive submerged slurries, ripping out streambanks, damaging bridges, and covering floodplain fields with gravel. Trees are ripped out and are swept away with the current. If they wash into a bridge or dam, they exert tremendous pressure, sometimes overcoming the forces holding the structures in place. Thus, the Hammond Bridge on the North Fork was no match for the many tons of woody debris pushed by the rushing current.

Main deck of the Hammond Bridge on the riverbank *(Springfield News Leader)*

Historic flood on the James River in May 2017 *(KY-3 News)*

Dams, naturally, are of special concern. Dams were built with maximum expected floods in mind, but those expectations, given the new wild card of climate change, may be out the window. Earthen dams, of course, are especially vulnerable. Once a berm is topped, a vigorous and rapid downcutting of rushing waters will quickly lead to its demise. Concrete dams may stand in floods, but earthen berms or wingwalls on one or both sides may blow out. Engineers will have to look at each dam individually to re-determine the magnitude of flooding event they can withstand. In many cases, dams will need to be beefed up or, if no longer needed, can be removed.

Streambank stabilization and protection practices may have to be reassessed in the light of future climate change-induced flooding. These practices include planting trees on streambanks or in floodplains, or using armoring, like big rocks or wing dikes, to protect banks. Like dams, these practices are designed to withstand certain levels of flooding. But if these levels are exceeded during climate-induced mega-floods, the hard work of restoration will have been for naught. With so many public dollars being spent these days on reclamation and restoration, this is an issue that will take a lot of hard thinking by resource agencies.

EPA tells us that over the last half-century, annual precipitation in the Midwest has increased 5-10%. But rainfall during the four wettest days of the year has increased 35%, and the amount of water flowing in streams during the worst flooding has increased 20%.[4] Intense rainstorms, like the one that moved over the North Fork watershed, will almost certainly cause increased damage and loss of infrastructure in the future. The flip side, unfortunately, is that warming temperatures mean hotter weather and more severe droughts may occur in between catastrophic floods. Springs will probably get wetter, late summers and falls hotter and drier.

Most parts of Missouri have warmed by an average of 1 to 1 ½ degrees over the last century. There are bound to be more extremely hot summer days, causing heat stress and health problems, especially for the elderly. Excessive heat will cause roads to buckle and crack. An overheated electric grid will be more susceptible to breakdowns, leading to more widespread blackouts. Longer and hotter droughts will mean more loss of crops. Higher evaporation rates will mean rivers flow for shorter periods, and reservoir levels will drop more quickly. During the drought of 2012-2013, public water supplies along the Missouri River were threatened when water levels dropped dangerously close to the lowest points of intake.

Scientists are hard at work to try to predict the kinds and magnitudes of the Earth's climate-induced changes, but it is largely a catch-up game. While studies are ongoing, the changes are happening, and rapidly. Growing up as a science nerd, I'm acquainted with the "scientific method." Experiments are set up to test a hypothesis, such as: if I add this chemical to a plant fertilizer, what do I expect the plant to do? I should make precise measurements of the plant's growth, for example, to see if the added chemical had an effect. In every experiment, I try to limit the variables, the factors that are manipulated or changed, to one or a very few. I want the same sunlight exposure, moisture conditions, and temperatures during every experiment so I can rest assured that those factors are not affecting the results. I try to isolate any changes to the effects of the variable I have introduced; in this case, the added chemical.

Rapidan Dam in Minnesota threatened by floodwaters, June 2024 *(ABC News)*

Climate change, by this line of reasoning, is a very complex experiment, indeed, with many variables interacting that could affect the specific outcomes. It is a novel experiment; one that has never been attempted in scale or scope—adding more and more carbon dioxide and other greenhouse gases to the atmosphere to see what will happen. The whole Earth is the test tube, the reaction chamber. We know the atmosphere will heat up, as it already has, and the oceans will warm, as they already have, and the icecaps will melt, as they are already starting to do. But beyond that, the intricate day-to-day effects are still largely a matter of conjecture, despite the ever-improving models. The bottom line is this: we are not sure what the ultimate results of the experiment will be, but we are taking an ill-advised gamble on what sort of planet we will be leaving for future generations.

The sad truth is that we've been expecting something like this to happen for at least sixty years. In 1957, in the science journal *Tellus*, the authors measured carbon-14 to carbon-12 ratios and determined carbon dioxide resided in the atmosphere for about ten years before being absorbed into the oceans. They warned that excess carbon dioxide could build up in the atmosphere and produce severe effects on climate and weather "throughout the Earth."

Hurricane Ian near Florida, September 2022, from the International Space Station (National Aeronautics and Space Administration)

In 1965, scientists on President Johnson's Science Advisory Committee predicted that "by the year 2000, the increase in atmospheric carbon dioxide will be close to 25%. This may be sufficient to produce measurable and perhaps marked changes in climate, and will almost certainly cause significant changes in the temperature and other properties of our stratosphere."[5]

These predictions have come true, despite the many naysayers. I remember a meeting in Arkansas in the early 2000s where a state senator confronted me with a question. We were standing by a window, with snow falling outside. Glancing out the window, he asked, "You don't really believe in all this global warming hooey, do you?" What do you say to that? In the first place, the question is oddly phrased, as if I may or may not "believe in" global warming, like I might or might not believe in Santa Claus, or the Tooth Fairy. I stumbled for an answer, trying to be diplomatic. "Well, the science seems to say we're heading in that direction." Not a satisfying answer for him, I'm sure, since scientists, after all, don't know everything and certainly don't all agree. It was easy for him to dismiss me, but at least not to my face, as another tree-hugger environmentalist shouting, "The sky is falling!"

Now, in 2024, it is difficult to deny that things are changing, and rapidly. Even legislators on both sides of the aisle are now admitting that climate change may be "real." Of course, they worry most about human lives and infrastructure—flooding will destroy roads and bridges, sweep houses and people away; hurricanes will destroy beachfront hotels and impair tourism; wildfires will destroy homes and displace populations. What politicians seldom discuss is the fact that the biggest and potentially worst effects of climate change will be on "nature," not directly on man and his built environment. Climate change is expected to damage ecosystems in new and profound ways.

Scientists have long theorized that climate change will exert its earliest effects on organisms occupying narrow niches—animals that live on only one kind of plant, for example, or inhabit places where living conditions, such as temperature variations, are very limited. One habitat type in that category is high mountains. An animal or plant adapted to live at high altitudes may have only a brief time to reproduce after snowmelt. As global temperatures warm, species may be able to gradually extend their ranges uphill, to cooler zones, but obviously, they can go no higher than the mountaintop. It takes years of data to establish how far or fast a species is moving, or for that matter, the specific effects of any number of potential weather variables, such as later or less snowfall. In Missouri and most other locations around the world, such long-term datasets are hard to come by.

One exception to this is at the Rocky Mountain Biological Station near Gothic, Colorado, where David Inoye has been recording plant and animal activity for over fifty years. He reports that compared to the 1970s, springs are now arriving earlier in the mountains. There is more rain in winter, sometimes melting snowpacks. Rapidly melting glaciers are increasingly spawning avalanches and rock slides. Snowpacks melt and shrink earlier in the season, and summers have become warmer and drier. High-mountain ecosystems, Inoye explains, are largely "driven by snow and when it melts." New conditions brought on by climate change have led to many "paradoxical consequences"—especially for plants that normally have only a few-week window to bloom after the start of snowmelt.[6]

Melting glacier
(National Geographic)

One of these consequences involves mountain daisies, which have started blooming earlier. This subjects them to the harmful effects of frost, which was formerly prevented by an insulating layer of snow. Frost damage reduces the amount of nectar in daisies, in turn making them less able to support a particular species of butterfly, which needs that food for egg-laying. Another example is the glacier lily, which now blooms seventeen days earlier than in the 1970s. Migrating hummingbirds have not kept up with these changes, now arriving after the lily's blooms have already withered. For other plants, the timing of seed production has changed, depriving small mammals and birds of critical food sources. Changes like these ripple through complex ecosystems, producing hosts of novel interactions and situations, many of them detrimental.

Inoye observes that ecosystems could "collapse" if global warming continues to "decouple" ecological relationships between plants and animals. "Multiple stressors are driving mountain ecosystems past tipping points," he warns. Massive disturbance events, such as the devastating bark beetle attack on mountains pines in the early 2000s, will happen more often. *Inside Climate News* co-founder Stacy Feldman points out that, "what people don't realize is that we're changing multiple things at the same time."[7] That's not a good scientific method.

With the Rocky Mountain dataset, we now have a better idea of what to expect as weather patterns change in the mountains. But what do we expect to see in Missouri? More intense floods and longer droughts, probably, and maybe more tornados, although "tornado alley" seems to have shifted southward. Perhaps more

bomb cyclones in winter, when cold and hyper-warm air masses are likely to collide, producing heavy rain; or more "atmospheric rivers" of moisture that go farther inland to dump rain on places that have historically seen very little. Without a doubt, multiple stressors will produce a complex mix of effects, possibly leading to certain types of "collapses" here as well. The effects may be subtle, at first, largely escaping detection, until the ball begins rolling down an ever-steepening hill, picking up momentum before its ugly crescendo, a devastating crash.

We can envision some simple examples of what could happen here. My favorite salamander, the hellbender, lives in cool, swiftly-flowing, highly oxygenated water. It needs a high concentration of oxygen dissolved in water because it "breathes" through its skin. As water warms, it rapidly loses its ability to absorb oxygen. A change of just a few degrees could mean the salamander can no longer get enough oxygen. A similar scenario could be posited for trout, which also require a high saturation of oxygen. Smallmouth bass, a favorite of anglers, also need cool, clear, highly oxygenated water. Could it be that smallmouth fisherman, like me, eventually have to leave Missouri to pursue their favorite fish?

Smallmouth bass prefer cool, clear water. *(Loring Bullard)*

Terrestrial animals and plants will also face new challenges. As in the mountains, Missouri wildflowers will probably bloom sooner, and possibly be bereft of nectar before the hummingbird calls. Trees that leaf out earlier in the spring may shade out the low growing plants under them that normally bloom in early spring, before the leaves on trees appear. Roadrunners, tarantulas, prickly pear cacti and evening primrose, denizens of hot, rocky glades, may expand their ranges northward as the terrain warms and dries. Cottonmouth snakes, which I have seen on rivers much more frequently over the past few years, may be creeping northward from their southern strongholds. Mosses, liverworts and ferns, secreted into cool alcoves of north-facing bluffs or along spring-fed streams, may have nowhere to go to survive rising temperatures.

So, what do we do about climate change? Conservationists now talk about climate resilience—ways to cope with and manage the negative impacts while trying to keep them from growing worse. That last part is the key. Any hope of stemming a dreadfully fearsome future rests on dialing down the production of heat-trapping emissions. For a "safe" climate future, the Union of Concerned Scientists (UCS) suggests that we must by 2030 cut greenhouse gas emissions by 50% from their 2005 levels.[8] This strategy is intended to address the primary cause of climate change, the burning of fossil fuels.

Mitigation is the term used to describe ways of doing this, such as using energy more efficiently and shifting to renewable energy sources, such as wind and solar. Environmentalists insist we must use less oil, ramp up public transportation, and use land in ways that remove, not add, carbon to the atmosphere. There is also a need for adaptation—preparing for and responding to the threats that are already here, such as flooding, wildfires, and hurricanes. The UCS suggests that solutions should be equitable. Climate impacts harm more people in certain vulnerable areas—such as older adults in cities, and residents of coasts and islands threatened by rising sea levels. As in most crises, the poor, the elderly, and people of color are more often in harm's way. Populations and communities at the greatest risk should receive priority for resources devoted to disaster planning.

We will need to make our landscapes more climate resilient—more able to resist the worst effects of climate change. Agriculture may need to shift its focus to more drought-resistant crops. Grazing lands could be converted from fescue to warm-season grasses, like big bluestem, Indian grass, and switchgrass, which have deep root systems to help them survive dry times. These grasses provide added benefits by storing more carbon than cool season grasses. We can manage forests better and help prevent wildfires by thinning trees and conducting prescribed burns, which remove flammable materials from the forest floor.[9] And we can add native plantings to our yards that are well-adapted to our climate, require less water, and that lock up more carbon than turf.

Environmentalists will need to keep up the pressure on the political front. The sooner the U.S. swings toward a "green economy," with jobs and communities devoted to renewable energy, energy efficiency, and energy justice—the better chance we have of stemming the flow of greenhouse gases into the atmosphere. So far, legislative attempts in this country to limit emissions from fossil fuels have been stranded on the shoals of political gridlock. But as conditions worsen, more politicians will concede something must be done. The situation, in many ways, is like the environmental crises of the 1960s. When enough citizens rise up in protest, march in the streets, and demand government action, the politicians will finally become motivated to act.

Somewhat ironically, I'm working on this chapter on the fourth of July, one of the hottest days of the year so far. It's about 93 degrees here—not all that bad compared to the high today in Sacramento, 111 degrees, but with Missouri's humidity, it's plenty uncomfortable. Our central air conditioning system runs almost constantly, pumping heat from our house and, unfortunately, adding to the local heat-island effect. My wife and I usually stay home on the July 4th holiday, rather than joining the travel madness and, at my stage of life, the obnoxious smoke and din of fireworks. But the news guys are reporting record numbers of vehicles on the roads, and a record number of passengers boarding airplanes. We surely don't act as if a climate crisis is pressing down on us.

Upon retirement, Beth and I looked forward to traveling—visiting foreign lands that for years have been on our bucket list. But journeys to faraway lands, though fun and enlightening, entail hopping aboard a big jetliner. Jet engines, guzzling high carbon aviation fuels, add a disproportionate share to climate change. I read recently that a plane trip from New York to London and back contributes as much greenhouse gas emissions as are produced in a whole year by a resident of Paraguay. I had to wonder, 'Why Paraguay?' but the point was clear. I had a decision to make—follow my retirement dreams, or live more sustainably. I still struggle with it. It won't be an easy decision to make.

How do we minimize the impacts of climate change on the earth, and especially on our critical life support systems, the diverse ecosystems that sustain us all? Ecosystems are already stressed, with biologically impoverished lands now spreading over much of the globe. Only about 5% of the land in the lower forty-eight states is in anything close to a "pristine, self-sustaining ecological condition."[10] Loss of diversity makes ecosystems more vulnerable to disturbances like climate change, with a smaller pool of potentially adaptive species available to weather the changes. Loss of biodiversity is both a cause of, and a result from, climate change.

Dust cloud over western Kansas, 1935
(National Drought Mitigation Center)

The answers, of course, are simple in principle, but difficult in practice. We simply must slow and eventually reverse the rising flow of greenhouse gases into the atmosphere. It would certainly help if government would ease and assist our transition to renewable energy sources. At the same time, we must work toward landscape resilience, our critical buffer against the worst effects of climate change. Largely, this is a matter of increasing biodiversity through regenerative agricultural practices and ecological improvements on the land. Every one of us can help by storing more carbon in our personal landscapes, for example, by growing more native plants. We must also put fewer greenhouse gases into the air—switching to electric vehicles, lawnmowers, and leaf-blowers; using public transit, whenever possible, or walking or riding a bike when we don't really have to drive, which will also benefit our personal health.

Hopefully, we are reaching a turning point in our recognition of the dangers of climate change. Many legislators, on both sides of the aisle, are at least discussing it. The military has already acknowledged that it presents a significant threat to world political stability. And as the detrimental effects of climate change-induced flooding, droughts, wildfires, tornados, and hurricanes reach even deeper into the regulation-wary Midwest, the conservative heart of America, people will realize that more must be done. It won't be easy, but even here, we may be approaching a new horizon, peeking over into a landscape of hope.

SOURCES

1 Southwest Missouri Council of Governments, *James River Watershed Management Plan*, 2023, p. 24

2 Insideclimatenews.org/news/07102019

3 Website, globalcarbonbudget.org

4 EPA.gov/19January2017/climate-change-mo

5 White House Report, *Restoring the Quality of Our Environment, Report of the Environmental Pollution Panel, President's Science Advisory Committee*, 1965; appendix Y4, "Atmospheric CO2"

6 Insideclimatenews.org

7 Ibid.

8 Union of Concerned Scientists website: ucsusa.org/resources

9 Megan Buchanon, personal interview, April 19, 2024

10 Douglas W. Tallamy, *Nature's Best Hope*, Portland, OR, 2019, p. 7

CHAPTER 12: LANDSCAPE OF HOPE

"When the land does well for its owner, and the owner does well by the land—when both end up better by reason of their partnership—then we have conservation. When one or the other grows poorer, either in substance, or in character, or in responsiveness to sun, wind and rain, then we have something else, and it is something we do not like."

ALDO LEOPOLD

My landscaping efforts at home are about hope, as much as anything else. With our pond, I hope to entice dragonflies, or even a few frogs or toads. The pond, however, seems better at attracting raccoons, which I'm sure enjoy catching and eating the poor goldfish. After making walls of stacked rocks, I hope to see lizards, but I fear the skulking neighborhood cats make life too hard for them. I want to attract birds, bats, bees, and butterflies as well, but I also hope to attract attention from my neighbors and friends, who I hope will consider making their own yards more "eco-friendly." My hope is tempered by the limitations of practicality, of course, but I hope enough lights will go on in enough brains to start making a little bit of difference in the declining quality of our biosphere.

Getting my yard in good shape, biologically speaking, is one thing, but what happens when I'm gone? What if the next owner tears out my precious native plants and replaces them with turfgrass, or fills in my tiny pond to make a planter for zinnias? There is no guarantee of what will happen to my "estate," of course, unless I secure its future through some sort of legal arrangement—like a trust, or an easement. It's something we all need to think about. Our conservation progress has been hard fought, and we don't want the gains we've made to slip away, whether on private or public land. There is just too much at stake.

I have several friends who have worked hard to make their land more productive, or attractive, or biologically rich. Some are leaving their land and improvements to heirs, who they hope will carry on their conservation legacies. Others are protecting their investments of time, money, and labor through instruments such as conservation easements. Some are willing their land to public institutions, with agreements on how the land will be managed after they're gone. In all these cases, the sustainability of land protection is the measure of success. Future generations, and the planet, will derive the benefits from their far-sighted stewardship.

This type of work on private lands is a critical adjunct to the public conservation efforts that reach back for generations—efforts that began with people like Theodore Roosevelt, who set aside some of our nation's most iconic landscapes. Later, his distant cousin, Franklin Delano Roosevelt, created millions of acres of wildlife sanctuaries. Today, with the enlightened leadership of many resource agencies and conservation organizations, the legacy of public land stewardship continues. Without these protected lands, our state would be hopelessly behind in efforts to preserve biodiversity and sustain ecosystems. It's that way nationally, too. America's national parks, grasslands, and monuments alone have safeguarded 85 million acres of land, an area twice the size of Missouri.

Missouri, like most states, has a long but checkered history of land protection. Wildlife protection came first, when we discovered, for example, that our rivers were getting fished out and our deer, turkey, prairie chickens and ducks were being slaughtered into oblivion. In the early 20th century, we passed game laws to halt these excesses. But these measures did nothing to protect the ecosystems where the threatened creatures lived. Next, we began to think about land preservation, but mainly the places we considered ruggedly beautiful or that would provide room for our fish and game to flourish. In the early twentieth century, game "preserves" came into vogue, often bought and managed by private hunting and fishing clubs.

Beginning in the late 19th century, we created National Parks. These magnificent landscapes were mostly in the west—places where the most egregious affronts of civilization had not yet reached. By the time Yellowstone Park was set aside in 1872, most of Missouri had already been plowed up, cut over, or overgrazed. As a result, the only places in Missouri eligible for national park status were small chunks of land with primarily historical, not ecological, values. Nevertheless, the national "best idea" stimulated many of Missouri's citizens to consider that our state had its own natural treasures that should be saved, somehow, from resource exploitation or development.

Missouri's extensive forest cover has certainly helped. About two-thirds of the state is forested. But by the first decade of the twentieth century, the state's forests had been heavily cut over, encouraging erosion and further impoverishing some of the "poorest" land in the state. This situation wasn't unique to Missouri, however. Forests all over the nation were in an alarmingly degraded condition, a potential threat to commerce and national security. In 1891, Congress passed the Land Reclamation Act,

Forests line Red Bridge Road in Christian County *(Loring Bullard)*

which allowed the president to set aside forest reserves. Forest management efforts increased in 1905 with the establishment of the Forest Service, an arm of the Department of Agriculture, under its first Chief, Gifford Pinchot.

The Weeks Act, passed by Congress in 1911, allowed the federal government to buy degraded forest land, an authority arising from the U.S. Constitution whereby land purchases could be made to conserve, redundantly, the "navigability of navigable waters."[1] Luckily, virtually all inland areas in the country are in the watersheds of navigable rivers, and all of Missouri's land drains to either the Missouri or the Mississippi. Over the succeeding decades, the federal government purchased about 1.5 million acres of land in 29 Missouri counties.

Collectively, these parcels became the Mark Twain National Forest, with a mission, as stated today, to "sustain the health, diversity and productivity" of the forests. In the past, however, the primary interest was on productivity, mainly logging. Although today's forests are managed more in line with the ideas of "health and diversity," management plans are still sometimes criticized by environmentalists as leaning too much toward lumber production and grazing, with not enough attention paid to ecosystem values.

Agriculture has long been a mainstay of Missouri's economy and the most widespread land-use, with 63% of the state's land, about 30 million acres, now in agricultural uses. About half of this is in crops, the other half in pastures and woodlots. Environmental protections on farmland have largely come through the efforts of conservation-minded farmers and tech-

Watershed of Bull Creek in southern Missouri *(Loring Bullard)*

nical and financial assistance from the federal government. Many land-conserving farming practices appear in cost-share programs, where the government pays some or most of the costs.

The idea, of course, is to incorporate environmental benefits into farming methods that also help the farmer's bottom line. But in practice, this is difficult, because farmers are naturally more concerned about profitability than conservation. They are often reluctant to enter contracts and encounter the dreaded government "red tape." County Soil and Water Districts, eager to make local farmers happy, may try to get them the practices they want, rather than what might be best for the land. As intended by Congress, the practices are always voluntary, and many farmers don't sign up. And for the practices that do get installed through cost-share assistance, there is little or no follow up after the contract period to make sure the practices are still working as intended.

Because so much of Missouri had been farmed, logged or mined by the end of the nineteenth century, state boosters, noting the public's increasing interest in outdoor recreation and tourism, promoted the idea that the state's most scenic natural areas needed permanent protection. In 1910, one of these boosters, Governor Herbert Hadley, organized a series of floating and fishing trips on the Current and White Rivers to call attention to the state's beauty and recreational potential. To create the state's first park, Hadley tried to push a bill through the state legislature to buy the Ha Ha Tonka estate, an amazing property near the Niangua River, for $16,000. The bill lost by one vote.[2]

A later Missouri governor, Arthur Hyde, in 1917 shepherded the first bill for a state park system through the legislature. A former car dealer, Hyde supported the "good roads" movement. Better roads, he reasoned, would make the state's scenic areas more accessible to the motoring public. The bill set aside 5% of fish and game license fees for a fund to purchase land for parks. But funds accumulated slowly, and while several parcels came up for sale and were desirable, such as Ha Ha Tonka, the state couldn't afford to buy. Hyde did manage in 1923 to get the assembly to increase the license percentage for parks from 5% to 25%.[3]

The next year, Hyde appointed Frank Wielandy, a major player in the Izaak Walton League, as the state Fish and Game Commissioner. Wielandy was a staunch supporter of parks, especially those with fishable waters, and by the end of 1924, the state had acquired or began acquisition of land for six parks. Four of them were at beautiful springs—Big, Round, Alley, and Bennett Spring—the others were Deer Run and Indian Trails. All these parks were already popular tourist destinations. By 1928, the state had fourteen parks under the administration of a state Park Board.

Missouri's natural resources, along with its people, suffered during the Depression years. When President Franklin Roosevelt set up the Civilian Conservation Corps in 1933, Missouri's citizens responded enthusiastically, contributing 25,000 young men to the nationwide effort. The CCC worked tirelessly on structural and ecological improvements, including the planting of over 3 billion trees. FDR did more for wildlife conservation than any previous president. By 1935, his administration had acquired over 1.5 million acres of land for forty-five new wildlife refuges, many of them near migratory flyways in the Midwest.[4] Missouri got a handful of federal wildlife refuges, including Mingo in the Bootheel; Swan Lake, Clarence Cannon and Loess Bluffs in north Missouri; and the Big Muddy complex, with separate parcels scattered along almost 400 miles of the Missouri River.

Wildlife refuges, national forests, and hunting preserves helped, but they were not enough. Early in the twentieth century, sportsmen called on the state government to step up and fill a void in resource protection for diminishing wildlife. In 1935, about 100 members of hunting and fishing clubs met in Columbia seeking to create a new agency to provide that oversight. They proposed an amendment to the state's constitution to create a new conservation agency. When voters approved the measure in 1936, 71% to 29%, it was one of the most lopsided votes ever on a constitutional amendment.[5] The Missouri Department of Conservation (MDC) was born, with a mission to conserve and regulate the state's "overstressed fisheries, forests and wildlife."

Many people suggested placing the state park system, created the decade before, under the MDC. As an apolitical body, the Conservation Commission could steer clear of the political strife that affected other state programs, including the park system. But the suggestion to place parks under the care of MDC created an awkward situation for MDC, because its primary responsibilities were fish, forests, and wildlife, not recreation or historic preservation. The issue of which governing body was best equipped to manage the state's parks wasn't totally resolved until sustainable funding for the park system became available.

Expansion of the state park system accelerated after 1964, when Congress created the Land and Water Conservation Fund (LWCF). The LWCF is funded from the sale of surplus land and taxes on motorboat fuel; and after 1968, from offshore oil leases. In 2020, the LWCF became a permanent part of the Great America Outdoors Act.[6] Missouri State Parks got nearly $18 million in LWCF funding between 1964 and 1966, providing upgrades and additions to fifty parks. These funds, however,

"Castle" overlooking Ha Ha Tonka Spring, ca. 1931 *(State Historical Society)*

Trout fishing at Meramec State Park *(Steve Spencer)*

Alley Spring, at one of Missouri's first six state parks *(Gayle Harper)*

were directed more toward recreation than toward conservation or preservation, and throughout the 1960s and 1970s, the park system had "relatively little focus on ecological matters."[7] But new directors in the mid to late 1970s, with the Parks Division by then under the Department of Natural Resources (DNR), worked toward increased ecological sensitivity.

Missourians heavily used and loved their state parks, and it became obvious in the 1970s that more funding would be needed for expansion and operations. In 1983, voters narrowly approved a constitutional amendment creating a 1/10 of 1 cent parks and soils sales tax, with funding to be split evenly between parks and soil and water conservation programs, both administered by DNR. The tax was renewed in 1988 by a two to one margin, again in 1996 by the same margin, and in 2006 by 71% in favor. The parks are now managed with ecological sustainability in mind. Over half of the parks in the system benefit from "stewardship practices" like controlled burns and invasive species removal.[8] Missouri now has 93 state parks and historic sites, a total of over 150,000 acres.

At times, being under the purview of DNR created problems for the park system. As a regulatory agency, DNR often encountered strained relationships with the Missouri legislature. Some legislators became upset with DNR over what they saw as "heavy-handed" regulatory actions on some of their constituents. For this and other reasons, there have been numerous attempts by the legislature to cut DNR's funding. Most of the legislators' ire was aimed at the department's Environmental Division, which regulates air, water, and land pollution, but there was always a risk of splash back on the Parks Division when the parent agency got punished.

The Conservation Department, created by a constitutional amendment, is less threatened by the whims of the legislature. It is also well-liked by Missourians, at least partly because its emphasis is on resources of direct interest to them—fish, forests, and wildlife. The cause of the organization was helped in 1971 with the launching of "Design for Conservation," supported by a 1/8 of 1 percent sales tax. Today, MDC administers about 975,000 acres in the state, 65% of which is forested, much of it in the dozens of Conservation Areas scattered around the state.

Missouri's Natural Areas Program, started in 1970, was intended to protect the best examples of the state's diverse ecosystems. Originally, the lands protected were all under the management of the MDC; forty-nine separate parcels by 1977. The Design for Conservation sales tax provided resources to protect additional areas. To help in this effort, DNR joined MDC to form an inter-agency team. Later, the Natural Areas Committee was joined by the Mark Twain National Forest, Ozark National Scenic Riverways, The Nature Conservancy, and U.S. Fish and Wildlife Service. Today, the program contains 180 designated natural areas, most of them in the south part of the state. These natural areas form the core of Conservation Opportunity Areas—key landscapes that receive focused conservation attention because of their high potential for conserving diverse habitats and the plants and animals that depend on them. The partners reach a broad consensus on where to get the most "bang for the buck."[9]

Despite its preponderance of farming, mining, forestry and development, Missouri had a few areas that could qualify as wilderness under the Wilderness Act passed by Congress in 1964. As early as the 1930s, Aldo Leopold had promoted the establishment of wilderness areas, especially to save iconic species like grizzly bears, wolves and mountain lions. "Like winds and sunsets," he said, "wild things were taken for granted until progress began to do away with them."[10] Wallace Stegner, literary hero of western wilderness, stressed that "something will have gone out of us as a people if we ever let the remaining wilderness be destroyed."[11]

But others looked at it differently. To some, the idea of a wilderness area "separate in every way from one's daily life," contradicted the notion that people are a part of nature.[12] Resource extractive industries saw it in more pragmatic terms. To them, wilderness designation essentially "locked up" the resources we all need. Even ecologists were not totally happy with the wilderness bill. Despite Leopold's influence on the wilderness movement, the Wilderness Act did not specifically protect ecological diversity or integrity, so "neither has been a dominant concern in the admission of new areas" to the system. Thus, the wilderness preservation system has been "less than a resounding success."[13]

Missouri's efforts to establish wilderness areas, like those around the country, were sometimes contentious, sometimes not. Most proposed wildernesses were contained within national forests and were already used extensively for hiking and wilderness camping, so there was little opposition to their designation. Some, like the huge Mingo "Swamp" in the Bootheel, were deemed "worthless" by many people anyway. But some, like the Irish Wilderness, were highly contested. Doe Run, Missouri's largest producer of lead, had set its sights on exploratory drilling in the Irish before it was designated. Only intervention by federal legislators saved it. The state ended up with a total of eight wilderness areas, providing a range of landscape features and ecosystem types, from dry glades on igneous mountaintops to vast wetlands with thousands of wading birds. The variety is impressive, but overall, the scale is not. Our 71,000 acres of wilderness in Missouri comprises less than two-tenths of one percent of the state's land area.

The protection of public lands in Missouri, about seven percent of the state, has made a huge difference. But about 93% of the state's land is privately owned, a major challenge for those seeking to expand conservation efforts. Of course, many private citizens provided land protection of one sort or another since before the state was formed. Some were guarding private hunting preserves or fishing areas. Some were overseeing sites of historical significance. But by the time of the establishment of Missouri's national forests in the early twentieth century, most of the privately held land had little in the way of conservation management. Few private lands were managed for ecological reasons until late in the twentieth century.

Dark sky night over a Missouri wilderness *(Ben Caruthers)*

Many landowners see conservation as something that should happen on public lands—something that won't "constrain the human activities we call progress." As Doug Tallamy puts it, "we cling to the notion that nature should be saved where nature remains, not where humans live, work and play."[14] Thankfully, interest on the part of private landowners, like the ones I know in Springfield, seems to be growing. One tool that is working successfully in many places is the conservation easement. With this tool, the landowner can continue to own the land but set part of it aside in perpetuity as an easement under an agreement recorded on the property deed. Thus, the easement rides with the property to every future landowner.

A continual challenge with using easements is that someone must be responsible for making sure the terms of the agreement are adhered to, also in perpetuity. Well-established land protection organizations, such as Ozark Land Trust, can take on this monitoring responsibility as long as funding, and with it, personnel, are there to support it. For this reason, the landowner may be asked to pay into a "stewardship fund" that covers up-front costs like legal fees and supports a growth fund to cover personnel for monitoring responsibilities.

Eroded bank of Huzzah River, before restoration work *(Rob Pulliam)*

The landowner may realize tax benefits from the easement, but these are often not enough to offset the required up-front payment. Therefore, easements are usually signed by conservation-minded landowners, not just by people looking for tax breaks.

Another land protection tool gaining traction is the "buy-protect-sell" model. This is as a viable way to put conservation on the ground through the private marketplace, obviating the need for government intervention. In one scenario, funds for the program are held by a bank or non-profit institution. This is referred to as "patient capital," because investor returns may be delayed as properties work their way through the process. A property comes up for sale—let's say ten acres. It has certain conservation values, so is of interest for protection. Maybe it has a creek running through it, but the creek has eroded banks and a thinly vegetated riparian area filled mostly with invasive species. A group or organization with a conservation mission uses the funds to purchase the property.

The group then leads efforts to fix the creek's problems—by stabilizing bank erosion, for example, or removing invasives and planting vegetation to expand the riparian zone. Then, a conservation easement is placed on the rehabilitated area, maybe three acres in all. The property can then be sold to a private buyer with the easement intact and recorded on the deed. The buyer, on the seven acres outside the easement, can develop, or farm, or build a retirement home, or whatever else local land-use laws will allow.

Money from the sale of the land is transferred back to the investment fund, ready for use as a "revolving fund" when the next desirable property becomes available. A corollary idea

Huzzah River streambanks after restoration work *(Rob Pulliam)*

is to have a second "philanthropy" fund, capitalized by conservation-minded people who just want to see land protected. This funding could pay for needed environmental improvements or be used to replenish the investment fund if the property declines in value with the easement attached.

The relationship between an owner and his or her land is a strong one and always has been. Resource agencies and conservation organizations must recognize that each landowner is unique, with his or her own set of concerns and priorities. Abigail Lambert, River Stewardship Manager for Ozark Land Trust, develops strong working relationships with landowners in the Huzzah Watershed. "The key thing about working with private landowners is to understand where they're coming from—understand their operation, and their needs," she said. There may be overlap of their needs with the organization's conservation values, and thus an opportunity for a "win-win," but, "We have to work—and especially listen—to find out what that is."[15]

Landowners tend to orient themselves to the land's future. What do I want this land to become? Who will take care of it when I'm gone? The more we think about the future, the more likely we are to want to shape our property in the image of "good land," a bountiful, beautiful landscape of fulfillment and contentment. Along the way, we hope, landowners will move up to the ethical sphere of management, as envisioned by Aldo Leopold; a land ethic idea that formed in his mind after many years of trying to improve the degraded condition of his Wisconsin farm.

Aldo Leopold *(Britannica)*

But the roots of Leopold's idea go back even further. In the 1930s, he visited the Sierra Madre Mountains of Mexico, an ecosystem then largely undisturbed by human hands. Here, he realized that, "all my life I had seen only sick land." It was a revelation to him. His eyes were opened to the "indivisibility of the earth—its soils, mountains, rivers, forests, climate, plants and animals," and he knew we must, "respect it collectively, not only as a useful servant, but as a living being."[16] This biotic view coalesced into his means of motivating others, not by instilling a fear of ecological catastrophe, but by leading them from "aesthetic appreciation to ecological understanding to love and respect."[17] These are the lessons we have yet to learn, fully, but we're making headway.

Many conservationists in Missouri are thinking outside the box, looking at new and innovative tools to get more conservation on the ground. One of these people is Megan Buchanon, Director of Resilient Lands for the Nature Conservancy in Missouri. A former employee of the Missouri Department of Conservation, Megan, while helping with the department's reorganization, realized her skill set included "connecting and integrating" land management science, philosophies, and people. I asked her why she seems so optimistic about landscape conservation in Missouri. Part of it, she says, is that there is now a "broad awareness" that more needs to be done. Not only is Leopold's land ethic idea more widely accepted, but there is also a growing "collaboration ethic" among agencies working to implement land-ethical conservation.[18]

Controlled burn at Woods Prairie in southwest Missouri *(Ozark Land Trust)*

Controlled burn at Woods Prairie in southwest Missouri *(Ozark Land Trust)*

As an example, she mentioned the "Habitat Strike Teams" that are fanning out across Missouri to do "boots on the ground" habitat improvements on both private and public land in the state's Conservation Opportunity Areas. Representatives of forty agencies and organizations put their heads together to develop the strike team concept. The teams, mostly young people, are learning conservation strategies as they perform prescribed burns or invasive species removal. In developing the program, the agencies put aside their logos and turf issues to find common ground, and to leverage the strengths of each individual partner. One agency might have funding available. Another might be able to contribute personnel, or equipment. It is a matter of "piecing the puzzle together" to form one cohesive team, working to repair and rehabilitate the landscape.

Megan is also about keeping the focus on fundamentals, the basic issues that "underlie our ability to solve all the problems." She calls it the "pothole" approach. Rather than getting into philosophical discussions about how to approach or solve the big problems, first seek an easy and quick consensus to fix a simple one—one that everyone understands, like filling a pothole. "There have been numerous examples where local and state politicians found bi-partisan support by side-stepping the polarized rhetoric and focusing on simple, shared issues with broad support," Megan said. Keep the work local, she advises, place-based, at the "human scale" where people live, work or play—and at the community level, where everyone has a shared ecological context. "That's how you can best foster everyone's personal land ethic—by focusing on connections between unique people and their own unique environment."

Being around a positive person always helps, because it is easy to become depressed about the future—about the terrible damages we are inflicting on our own homes, our biological partners, our biosphere. It is harder to have hope that we can rise above our more selfish and uncaring natures. But it is hope that makes us want to connect more with the Earth and its myriad inhabitants; and in them, we also connect to each other. No matter how bad things get, hope can get us through. It drives us to not only better ourselves, but to improve the lives of everyone in the future.

It is good to know that we have people like Megan Buchanon out there to bring us new ways of thinking, new ecological insights, and powerful collaboration. And people like Doug Doughty, the farmer in north Missouri, who can see a better, more sustainable path forward for Missouri agriculture. And my friends, Bob and Barb Kipfer, who unselfishly share their land and knowledge to impart Leopold-style ethics to future generations. These powerful connections of people to their land, and people to each other, gives me hope that the strong web of shared values will be continually enlarged, spreading a message of hope and inspiration to those who want a better world for us all.

SOURCES

1 Jane B. Snow, "Implementing the Weeks Act: A Lawyer's Perspective," *Forest History Today*, 2011, p. 70

2 Missouri Life Inc., *Missouri State Parks and Historic Sites: Exploring Our Legacy*, Susan Flader, Ed., 2nd edition, 2016, p. 2

3 Ibid., p. 3

4 Douglas Brinkley, *Rightful Heritage: Franklin D. Roosevelt and the Land of America*, New York, 2016, p. 297

5 MDC.mo.gov/magazines/conservationist/2011-11/missouris-unique-conservation-legacy

6 Douglas Brinkley, *Silent Spring Revolution*, New York, 2022, p. 460

7 *Missouri State Parks and Historic Sites*, p. 14

8 Ibid., p. 24

9 Mike Currier, personal interview, July 22, 2024

10 Soil and Water Conservation Society: *Aldo Leopold: The Man and His Legacy*, Thomas Tanner, Ed., 1995, p. 25

11 *Silent Spring Revolution*, p. 185

12 Ibid., p. 399

13 *Aldo Leopold: The Man and His Legacy*, p. 35

14 Douglas W. Tallamy, *Nature's Best Hope*, Portland, OR, 2019, p. 20

15 Abigail Lambert, personal interview, July 31, 2024

16 *Aldo Leopold: The Man and His Legacy*, p. 9

17 Ibid., p. 23

18 Megan Buchanon, personal interview, April 19, 2024

EPILOGUE

At this stage of my life, I am no longer embarrassed to call myself an environmentalist—a tree-hugger. A lifetime of working toward a healthier environment has convinced me of its critical importance to the quality of life we have come to enjoy and expect. Protecting the planet, our home, is really self-preservation. We must all think more ecologically if there is any hope of achieving a sustainable future, where the wants and needs of future generations can be met. We must all become environmentalists, to some extent, if we care about the quality of lives that our kids, and grandkids, and future generations will have. Thinking in an ecological context, moreover, makes it easier to understand, and guide, human behavior—to bring awareness of how our myriad connections with the natural world sustain us, and allow us to thrive.

We all connect with nature every single day, if we stop to think about it. But of course, most of us rarely think about it. As John Muir once observed, "most people are on the world, not in it." I believe that many people have felt a degree of emptiness inside, not realizing they have lost their firm footing in the Earth, as opposed to on it. Most people will say they are concerned about the quality of the environment. Most have at least a kernel of a conservation ethic. The job of the environmentalist today is not to beat people up over past behaviors, but to gently guide them back into the Earth—help them re-establish, and re-affirm, their natural roots.

People vary widely in their levels of optimism, or lack of it, when it comes to the environment. Some die-hard environmentalists are doomsday prophets, like the guy in the Li'l Abner cartoons who walked around under a black rain cloud—the people we get tired of being around, and shun at parties. Some predict a coming ecological apocalypse, the imminent demise of the human species. I don't believe that for a minute. Humans, which are at least as adaptive and resilient as rats and cockroaches, will be on this planet for many more thousands, maybe millions, of years, barring an asteroid impact or all-out nuclear war. The question is not whether people will be here in the distant future, but what life will be like for them.

There is no way to know, of course. But I do believe we are approaching a fork in the road. One road leads us into a downward spiral, a quickening degradation of land, water, air and ecological health, and, as a tragic corollary, an ever-diminishing quality of life. The other leads to human enlightenment and thoughtful stewardship—more ethical behavior toward the planet, and each other. Of course, "fork," as a metaphor, is a vast oversimplification of how things will almost certainly play out. We will never follow a single road to an imagined, idyllic future. We will travel a maze of winding routes, zig-zagging, sometimes back-tracking, running in fits and starts.

People will follow widely divergent paths. Some, safe and snug in their comfortable existence, will build empathy walls allowing them to ignore the suffering of their fellow travelers. Others will barricade themselves from the rest of the world, retreating to their well-stocked bunkers. The rest of us will no doubt muddle through as best we can; trying to adapt to changing conditions and more than likely, search for health and happiness in a biologically impoverished world. Some of the coming changes have the potential to bring large measures of human suffering. Hopefully, in the face of this, we will not lose sight of our humanity. We will render what aid we can to the less fortunate and to the extent we can, work together to make life better for everyone.

These days, social and moral issues seem to have taken precedence over environmental ones. Though we may be wide apart on some issues, most of us, I think, want the same basic things. We want our children and grandchildren to inherit a world that is healthy and safe. We don't want them to breathe polluted air, or eat food or drink water tainted with bacteria or poisonous chemicals. We want them to enjoy beautiful scenery and be able to visit, if not live in, biologically rich landscapes. No matter where we are in our personal journeys, there should be a wide expanse of common ground between us, a natural meeting place of the minds.

Before we can set foot on a more sustainable path, however, there must be a realization that things could be better. Many people think our worst environmental problems are far behind us. They take the quality of today's environment for granted—one of their "rights" as an American. They don't appreciate the struggles and hard work that have allowed them the freedom to not worry about it. They have never learned the simple truth that their quality of life is almost entirely dependent on a high quality of environment. The threats they see are not in the natural world, but in other people who would make their lives more difficult, or dangerous.

For many of us, the "new" environmental threats may be harder to recognize. In the 1960s, the threats were direct, personal, and highly visible—everyone could see the hazy, polluted air, or smell the waste dumps, or see rivers that "oozed" rather than flowed. The climate and biodiversity crises of today are less direct and less visible in the day-to-day experience. Although rapid in geologic time, the disappearance of species is much less obvious from year to year. Hurricanes, wildfires and floods, though highly visible, are, unlike pollution, widely considered "natural" events. With this as a mindset, larger and more frequent hurricanes, or floods, or wildfires, can be more easily dismissed as aberrations, large but incremental straying from the natural, historic norm.

Today's environmental crises will no doubt force change upon us. In some areas, transitions are already occurring. Slowly but steadily, we are moving away from a reliance on fossil fuels, the primary source of greenhouse gases. Shrinking reserves and rising prices of coal and oil will ultimately make renewables not only more attractive, but necessary. Electric cars and lawnmowers will someday be ubiquitous. There are already a lot more people working in the solar industry than in the coal industry. Utility companies are busy trying to figure out how to phase out coal-fired power plants, with their expensive fixes for greenhouse gas and toxic coal ash problems. The transition to a "greener" economy is already underway, and will accelerate if the leaders of the country will allow and assist it.

In other areas, change is coming slowly. For example, the biodiversity crisis is barely on the popular radar, although most people intuitively sense that losing species can eventually lead to human harm. Even here, a path forward is opening for us, if we would just pay attention. Doug Tallamy suggests the creation of a "Homegrown National Park." If enough private landowners unite in their efforts to create biologically rich yards and rural landscapes, we would produce, in total, more ecologically healthy land than in all the existing National Parks combined. And, as Tallamy asserts, ecological restoration at home can bring great satisfaction. Each of us has a role to play. We can talk with our friends and neighbors about small steps toward not only thoughtful stewardship, but also the happiness and fulfillment that comes with it.

Government can and should play a leading role in the environmental reawakening. When most politicians finally realize that a healthy economy is utterly dependent on a sustainable relationship with the Earth and its resources, they will push for more eco-centric policies, just as they did in the early 1970s. And as in all profound social transitions, there will be two required elements—the grassroots piece, as exemplified on the first Earth Day, when a rising tide of concerned citizens demanded change—and powerful, enlightened leadership; especially leaders who can orchestrate efficient and fair mechanisms of change.

The words of Aldo Leopold still resonate from the early days of the environmental era. He sought to change the culture that viewed land as merely something to be used—to be bought or sold; to have its resources extracted, or to be made artificially productive. While he recognized land as a great servant, Leopold advocated a more sophisticated, ecological approach to land as a place where someone could build a good life. Perhaps his greatest gift was his ability to lead people from aesthetic appreciation of land to managing it through ecological enlightenment, and from there, eventually, to love and respect. Those are things the land, and the world, desperately need.

Reaching people with a message of urgency, while holding out hope, is a delicate balancing act. Think about the land ethically, Leopold advises. Start small, Tallamy suggests. Work on simple problems with simple solutions first, Buchanon advises—the "pothole" approach. We can't be afraid or embarrassed to take those first small steps. It is relatively easy to get people to plant natives that will attract birds and butterflies to their yard. It isn't hard to get them to buy locally produced meat, instead of supporting industrial-scale animal agriculture, or buy an electric vehicle or lawnmower, as long as they're easy to use and cheap to operate. Simple steps like these can build to an awareness that change needn't be painful, but can be easy, even pleasing. Small changes can make life less stressful, and maybe even more beautiful.

There is much work to be done if we want to make life better for future generations. It will be challenging, but Americans have always been good at rising to the occasion; coming together at critical times; taking a leadership role. Being a leader means asking the difficult, but essential, questions: How do we fulfill our ethical responsibilities to our neighbors, our communities; to the rest of humanity, and to the planet that sustains us? How do we make the world a better place, not just for ourselves, but for the potentially infinite number of people in the future? If we honestly search for answers to these questions, we can surely find plenty of common ground with people who don't agree with us about many other things.

I want America to show the world a more sustainable path into the future. I want my grandchildren and their grandchildren to inhabit a healthy, happy, culturally and biologically rich world. I believe we are up to the task. But if we are to lead, we must remember, and take to heart, the lessons of the past. We must honor, respect and acknowledge the times we worked together, stood firm, united, in the face of seemingly insurmountable challenges. As Americans, we can take pride in the fact that we, the free and the brave, have long served, and can continue to serve, as a "light unto the world." I will not be here to see it, but I will always have hope that the light will not only shine undiminished into the coming centuries, but will burn brighter still.

ABOUT THE AUTHOR

For twenty-three years, Loring served as Executive Director for the Watershed Committee of the Ozarks, a non-profit organization working to protect sources of public drinking water for Springfield and Greene County, Missouri. For thirteen years, he worked in environmental health, and now serves as a watershed consultant for the James River Basin Partnership. He taught environmental science and aquatic biology at Drury University, and has a Master's degree in Natural and Applied Science from Missouri State University. Loring currently serves on the boards of two non-profit organizations, Ozark Land Trust and DarkSky Missouri. He has published five water-related books: *Living Waters: The Springs of Missouri* and *Saving Bull Creek* for the Ozark Studies Institute of Missouri State University; *Consider the Source: A History of the Springfield Public Water Supply*, for the Watershed Press; *Healing Waters: Missouri's Historical Mineral Springs and Spas*, for the University of Missouri Press; and *Springs of Greene County* for the Missouri Department of Natural Resources, as well as numerous articles for *Missouri Historical Review*, *Missouri Conservationist*, and *Missouri Resources*. He has a lifelong love of water and enjoys canoeing and fishing on Ozark streams. Loring is married to Beth, has two grown daughters, six grandchildren, and lives in Springfield.

INDEX

www.ingramcontent.com/pod-product-compliance
Lightning Source LLC
LaVergne TN
LVHW080047210926
848481LV00011B/3295
* 9 7 9 8 9 9 4 7 7 2 1 1 9 *